WITHDRAWN

MICROFILM

Tomorrow's Transportation: Changing Cities, Economies, and Lives

For a listing of recent titles in the *Artech House ITS Library,*
turn to the back of this book.

Tomorrow's Transportation: Changing Cities, Economies, and Lives

William L. Garrison
Jerry D. Ward

Artech House
Boston • London

Library of Congress Cataloging-in-Publication Data
Garrison, William L.
    Tomorrow's transportation : changing cities, economies, and lives / William L.
Garrison, Jerry D. Ward.
        p.   cm. — (Artech House ITS library)
    Includes bibliographical references (p.) and index.
    ISBN 1-58053-096-6 (alk. paper)
    1. Transportation—Forecasting.   2. Technological innovations.
I. Ward, Jerry D.   II. Title.   III. Series.
    HE147.5.G37   2000
    388.4—dc21   99-058763
                                                                          CIP

British Library Cataloguing in Publication Data
Garrison, William L.
    Tomorrow's transportation : changing cities, economies, and
    lives. — (Artech House ITS library)
    1. Transportation—Forecasting   2. Transportation—
    Technological innovations   3. Urban transportation   4. Traffic
    engineering—Technological innovations   5. Intelligent
    control systems
    I. Title   II. Ward, Jerry D.
    388'.0112
    ISBN 1-58053-096-6

Cover design by Igor Valdman

© 2000 Artech House, Inc.
685 Canton Street
Norwood, MA 02062

All rights reserved. Printed and bound in the United States of America. No part of this book
may be reproduced or utilized in any form or by any means, electronic or mechanical, including photocopying, recording, or by any information storage and retrieval system, without permission in writing from the publisher.

All terms mentioned in this book that are known to be trademarks or service marks have
been appropriately capitalized. Artech House cannot attest to the accuracy of this information. Use of a term in this book should not be regarded as affecting the validity of any trade-
mark or service mark.

International Standard Book Number: 1-58053-096-6
Library of Congress Catalog Card Number: 99-058763

10 9 8 7 6 5 4 3 2 1

Contents

Preface

Transportation has threaded its way, sometimes invisibly, through the fabric of our world. Nearly all the productive or social processes of working and living involve transportation, either as an integral part of the process itself or in the activities on their periphery. Better transportation enables improvement in almost all these processes, often dramatically and often in ways not now imagined. So advancements in transportation can make our collective future better over time in ways that transcend the transportation itself.

Our experience tells us this is a modest assertion; it is exactly what transportation improvements have done over and over in the past.

So our central thesis is that better transportation leads to a better world. We anticipate and encourage improvements not because it's more fun or more convenient to go faster or travel more cheaply or ship more goods, but because we know such improvements extend to many other aspects of our lives.

The last great waves of change in transportation produced the railroads, the steamship, the auto and the truck, and finally the air system. Most of these began before many of us were born and are now an accepted part of today's scenery. The revolution that produced them, largely reflecting the mechanical arts of the 1800s, ran its course decades ago.

Today's changes are coming from a new revolution. While we have had a foretaste with the telegraph, radio, and telephone, we are almost surprised daily, and frequently delighted by, progressive innovations in communication, computation, and means of sensing the environment around us. We also have lighter and stronger materials and more versatile and efficient sources of power. These technological building blocks bring possibilities in both transportation and communication that are almost beyond imaginations still geared to even a fairly recent past.

Some of these new possibilities are already on stage, some are still waiting in the wings. Some represent entirely new capabilities, and some are variants on existing systems. Many are embodied in the Intelligent Transportation System program, which has been designed to exploit new technology in improving the nation's road and highway system. We hesitate to call all these possibilities predictions, although we believe that many of them will, in fact, come to pass.

In exploring their likelihood and their possible impacts we have tried to be consistent with evolving social, economic, and ecological realities, while recognizing concomitant advancements in communication and other relevant technologies.

We clearly have our problems. Our growing suburbs (and exurbs) eat up land, with all its implications for wildlife and natural ambiance. Growth in our roads has not kept up with the need for them, creating congestion that is a real thorn in the side of many commuters and an added cost to nearly all commercial activity. We cannot stop growth, but we would surely like to make it smart growth. And we think that part and parcel of smart growth is versatile and abundant transportation.

The newspapers, the politicians, and the social critics tell us almost every day what ought to be done. Some proposals ask that we return to living as things were in an earlier age: the centerpiece of their prescription is usually to get rid of the automobile, limit urban development, and build transit and intercity rail systems. We are told that problems would go away if there were high tolls on roads and highways, or if gasoline prices were at the level they are in Europe or Japan. Sometimes bike paths are part of the solution. There are lots of ideas about private sector financing and privatization.

We have tried to sort through all these prescriptions and prognostications and hone in on those things that we think make sufficient sense to really happen. We lay out our lines of reasoning for all to see so that you, the reader, can judge their merit through your own vision.

The pressures motivating change derive from both the advent and recognition of new technical capabilities and new alternatives, from reactions to obvious dysfunctions in existing systems, and from shifting social attitudes and values. But change is always in tension with the status quo, the existing ways of thinking and working, as well as with the investments already made in older technologies. So just identifying what appears to be desirable directions of change does not necessarily make them happen. So we pay considerable attention to the recipes for change, the conditions and circumstances that appear to be necessary to bring it about.

In Henry IV, Part I, Shakespeare showed that he understood the problem:

> Glendower: *I can call the spirits from the vasty deep.*
> Hotspur: *Why so can I, or so can any man: But will they come when you do call them?*

We have striven to identify possibilities and the paths along which they might come when and if society calls.

Our first chapter lays out a bit of historical perspective and makes some observations about ways of thinking about transportation. We then turn to the future and focus in the next five chapters on our dominant mode of personal transportation: the car. We start by presenting the apparently ridiculous notion of cars that will be able to drive themselves. It isn't really ridiculous: the first technical steps toward this almost unbelievable turn of affairs are, in fact, upon us now. The implications for the future are profound.

Next we address the almost ubiquitous problem of traffic congestion and what might be done about it, summarizing with a discussion of the general issues of personal mobility in our cities. In the last chapter of the section, we discuss the possibility of new kinds of automobiles.

The next section of the book is devoted to cities and how they might change in the future with the advent of new kinds of transportation to serve them. The modern city is vastly different from those of a century ago, and we have no reason to think that the city of a few decades from now won't be very different from those of today. We discuss both past evolution and some of the possible paths for its continuing into the future. Cities are still—and perhaps always will be—very much a work in process.

Any discussion of transportation must include energy and environmental considerations. After looking at the numbers that help define the issues, we point out some options that may aid continued progress toward more energy efficient and less polluting transportation.

The line between urban and intercity transportation is becoming increasingly blurred, particularly when one's perspective is less on spatial arrangements and more on the economic activity of the city and the personal travel habits of its inhabitants. We begin our discussion of intercity transportation with the railroads and possible futures for them. We then look at a possible direction of evolution of our highway system. Last we treat some completely new high-speed ground systems. We introduce this group of chapters by describing a bit more fully the birth of the railroads, which is a classic illustration of the process of invention and innovation.

The next section treats the air system. The first chapter of the section is a kind of generic exploration of possible new ways to structure our commercial airport system.

The second chapter is a notional case study of the Los Angeles Airport System. At the time this book is being written, there is being debated a very ambitious and expensive plan for expansion of LAX, the primary airport of the Greater Los Angeles Metroplex. We step into this specific issue, applying some of the thinking we had presented earlier, by playing the role of a very long-lived planner who looks both forward from LAX's birth in 1928 and back from the year 2020 to tell us how the issues of 2000 were worked out.

In the last section of the book we discuss the synergism between new developments in communication and the transportation system. We close with a few observations about the future.

The reader should know a bit about the authors. We are technological optimists, but consider ourselves pragmatic fellows. We proceed with hope rather than cynicism. Our priorities go to new ways to provide services rather than to policy development or revision. We present ideas in the spirit that they just might be useful, which is for the future to determine.

And we want the reader to know of those who have aided us. We appreciate the many thoughtful comments and suggestions we have received on our progress, and cite in particular, our constructive critics Forrest R. Pitts, David Gillen, and William Spreitzer. Melvin W. Webber taught us much about presenting and playing with ideas. We also thank Dr. John McCarty of El Segundo for urging us to write the chapter on the Los Angeles Airport System, for generously providing us access to his extensive files on the project, and for his very useful insights. We, of course, accept all responsibility.

We hope you find this book interesting and, in a few places, at least mildly entertaining.

1

A Bit of Perspective

Without a long running start in history, we shall not have the momentum needed … to take a sufficiently bold leap into the future.

Lewis Mumford, *The City in History*

In 1712, not quite 300 years ago, an Englishman named Thomas Newcomen (1663–1729) put the first steam engine into commercial service. It was a vertical cylinder in which a piston was pulled up by a counterweight, steam was introduced to displace the air under the piston, and when the steam condensed the pressure of the atmosphere forced the piston back down, doing work. These "atmospheric" engines were used to power pumps to remove water from coal mines.

These engines were heavy, slow, and so inefficient that it's hard to believe today that they were worth the trouble. But the door Mr.

Newcomen was opening was wider than he could have ever imagined. As it had forever, Mr. Newcomen's world ran on muscle, supplemented only by wind in sails and occasionally the force of gravity as it acted on water. In Mr. Newcomen's world no one had ever traveled faster than a horse could carry them.

Many people give credit for the idea of the atmospheric engine to a Huguenot Doctor of Medicine and Professor of Mathematics, Dr. Denis Papin. Dr. Papin had done work on the preservation of foods using steam and had also tried to build a gunpowder engine, so it is not implausible that he did, in fact, come up with the original notion—some 20 years ahead of Mr. Newcomen. But unless we happen to be ancestors of one or the other of these gentlemen, settling this controversy is not terribly important to us today. We shall give them both credit, as many historians of technology do [1].

It took another 50 years after Newcomen's first engines—more than an average man's lifetime in those days—for the next big step forward. An instrument repairman at the University of Glasgow named James Watt (1736–1819) was repairing a model of the Newcomen engine and thought he could do better. With no help whatsoever from a tea kettle, he began a series of improvements that led to the first real steam engines. Watt's first patent was some sixty years after Mr. Newcomen's seminal step [2].

By the early part of the nineteenth century, these heavy, awkward, dangerous devices had evolved to bring us the railroad and the steam-ship—and the Industrial Revolution.

A new attitude had been nurtured in the western world in the prior century: a growing democratization of minds, a greater tolerance toward change, the separation of science from theology and antiquity, a willing-ness to experiment. These all led to a new vigor in technological advance. There's little doubt that the steam engine, this first substitution of machine power for human and animal muscle, was a key event in this progression, a wavelet that by the nineteenth century became a tide.

It is hard to appreciate today the magnitude of the change wrought by these seminal steps. Man's earliest transportation was his own two feet. Very early on, various beasts of burden, including horses, entered the picture, as did the boat and the ship. The early impact of the wheel is overestimated today because wheels need roads and good roads are hard to build and even harder to maintain. Not until the Romans, were wheeled vehicles other than chariots used extensively, and most of their roads were designed for walking. Before the nineteenth century the best roads were those built by the Romans and they were designed for walking, not for wheeled vehicles [3].

For thousands of those ancient and not-so-ancient years, gains in capability were a succession of small, incremental improvements to these basic systems: beasts of burden, animal-powered vehicles, and sailing ships. It's hard for us today to get excited about these kinds of improvements. From our modern vantage point—cruelly out of context—we cannot appreciate that such early inventions as the stirrup or the horse-collar were great leaps forward in their day. The stirrup made horseback a less precarious mode of transport and let a man swing a sword without falling. The horse-collar increased the pulling power of this most speedy of domestic animals by probably a factor of five. That is not hay. As the evolution of roads illustrates, much less consequential changes than these are the kinds of advances that typified thousands and thousands of years of transportation evolution [4].

Then came the steam engine. And in the last two hundred years we have added the train, the steamship, the automobile and truck, the airplane—and ubiquitous, well-surfaced roads. We will tell more of these stories later.

These new forms of transportation—vast improvements over systems powered by muscle and sail—were not just wonderful new things evolving alongside all the other marvels of this revolution in the western world's economies, they enabled many of these marvels to happen. Because transportation is such an integral part of nearly all commercial and social processes, transportation improvements set the stage for much more widespread innovation in the nature of these processes.

Better transport increases the size of the markets available to factories, and larger markets let them expand their scales of output. Expanded scale encourages increased productivity because it makes economical more labor-saving machinery and greater specialization. Trucks and trains let factories move to cheaper land. More land permitted better physical arrangements for factory operations.

Like pebbles in a pond, the new forms of power and the resultant innovations in our transportation had effects that rippled outward to enable change in the spatial arrangement of the our cities, the nature of interactions within and between them, the design and placement of our factories and our farms, the distribution of goods, and the scope of our social interactions. Note that we say "enabled," not "caused": transportation in itself does not cause change, its improvement widens the scope for other motivations to be realized. But the result has added up to profound change in the world's economy, society, and polity.

By the beginning of the twentieth century ships had shrunk the oceans by a factor of four or five in time and permitted routine interaction among continents. Railroads let cities grow where there was no water transportation. Agriculture could move to the best land and climate.

Once the primary transport for large goods movements was waterborne: the ocean-going ship and boats and barges on rivers and canals. Then we added the railroads and pipelines; now the truck competes with the car for space on our roads and highways.

The airplane has shrunk the world. It has created a level of personal and commercial interaction over distances almost unimaginable to our great grandparents. We now routinely eat grapes from Chile and fish from Australia.

The average citizen does not see all of these impacts: the movement of goods that keep our factories working, food in the supermarkets, and a diversity of goods that almost defies belief in our stores.

Whether applauded or deplored, new forms of urban transportation have let our cities reshape themselves. First the trolley and then the automobile have widened the options for home location and created the possibility of entirely new spatial arrangements in our cities. The personal reach of the average citizen has widened dramatically.

In changing, though, the cities have outgrown and compromised the kinds of transportation that originally enabled their metamorphosis. We have become automobile dependent because the low-density city that the automobile has made possible is difficult to serve with the standard forms of public transportation. Traffic congestion is a severe problem in our larger cities, and air quality in many.

The window on the world for most people is their car window

The average citizen senses, but usually underestimates, the degree to which the automobile, along with small trucks and vans, have come to dominate personal transportation. In 1994 roughly 83% of passenger-miles traveled was in these personal vehicles, of which only a tiny portion was in taxis. Drivers and other riders in larger trucks bring this up to 87%. Commercial air accounted for 9% and school buses for 2%. Transit, intercity bus, motorcycles, private aircraft, and Amtrak together made up the last 2% [5].

Where the industrial revolution added power to the world, the current revolution is adding smarts. A deeper knowledge and a rapidly increasing practical understanding of solid-state physical phenomena are bringing us daily new and astonishing capabilities in data manipulation and computation, communication, and sensing.

These new capabilities are being applied to produce smarter traffic control, smarter highway signs, ship and railroad signal and control systems, aircraft and air traffic control, and toasters and televisions. Locomotives, for example, are being fitted with on-board track signal systems, electric train braking, and wheel slip control for better traction. The hype that accompanies the introduction of these new capabilities not withstanding, there is real progress taking place that we think will astound us all long before that oft-cited "long run."

And so...

Waves of transportation developments have enabled (1) larger markets that (2) opened opportunities for greater specialization, larger capital investments, and for process innovation in general, (3) thus giving us new ways to do old things better as well as entirely new things. Better personal transportation has widened the scope of social, commercial, and institutional interaction. We reiterate that the key word is enabled, not caused: transportation improvements do not cause things to happen but make it possible for them to happen in response to other motivations.

We have come a long way along a not-entirely-smooth road. But we are far from having gone as far as we can go; there is still a need and an opportunity to exploit the opportunities that new technology and new understanding have brought us. We can go further to continue innovation in our transport systems, not just to alleviate the rough spots and obvious deficiencies in what we have, but to add entirely new capabilities.

Change will happen and probably without our even being aware that the process is underway. Remember that change is deceptive. It lurks in the shadow of the status quo, looking either innocuous or improbable, and only comes out into the open through what often looks to be accidental or unusual circumstances or long after the seminal step has been taken. Rarely does anyone foresee the downstream consequences of such seemingly innocent or only half-formed beginnings.

This book is about change, some of the opportunities in our future, especially the ones that could really happen. Some are already on stage; some are still waiting in the wings. We think many offer the promise of more pleasant and exciting environments for living.

Let us begin our story with the car that can drive itself.

References

[1] Derry, T. K., and T. I. Williams, *A Short History of Technology*, London: Oxford University Press, 1960.

[2] Dickinson, H. W., *A Short History of the Steam Engine*, New York, NY: The MacMillan Co., London, Cambridge University Press, 1939, p. 68ff.

[3] Leighton, A. C., *Transport and Communication in Early Medieval Europe: AD 500–1100*, London: Newton Abbot, David and Charles, 1972, pp. 51,74.

[4] Lay, M. G., *Ways of the World: A History of the World's Roads and of the Vehicles That Used Them*, New Brunswick, NJ: Rutgers University Press, 1996.

[5] Calculated from data in *National Transportation Statistics 1996*, Washington, D.C., Bureau of Transportation Statistics, U.S. Dept. of Transportation, 1995.

Part I

The Car and Traffic

2

The Car that Can Drive Itself

The car that can drive itself? Ridiculous!

The Model-T Ford wasn't the first automobile. In the 20 years prior to its introduction in 1908, there had been many wondrous and varied gasoline-, steam-, and electric-powered automobiles, including four earlier models of Fords [1].

But the Model-T was the first automobile specifically designed for what was then high-rate production and intended to be priced within the reach of the average family. By 1912, annual production had reached 78,000; and by 1916 it was up to a half million. Before the Model-T was superseded by the Model-A in 1927, some 10 million had been built and sold.

The first giant step toward automation was the addition of a battery and self-starter in 1919. Before this, starting the car was an adventure in itself, as we shall briefly illustrate.

If you're the prudent type, start by sticking a ruler in the gas tank to check the fuel level. Then get in the car and pull one of the levers on the steering column; this sets the spark to "retard." The other lever on the column is the throttle; set it about half way. Next pinch the handles on the emergency brake, which is also part of the shifting mechanism, and pull it all the way back.

Now get out and go to the front of the car. (You say it's raining? You have our sympathy.) First pull out the choke; that's the little loop of wire on the left side of the radiator. At this point it often helps to turn over the engine a few times to draw a little gas into the cylinders; you do this using the hand crank that sticks out of the front of the engine.

Back to the driver's seat and set the spark to the position for starting. You're ready! Now to the front of the car again, grab the hand crank, and pull up smartly.

OK, so it didn't start that time. But it didn't kick back and break your arm, so keep trying.

It starts! Now run back and advance the spark some more and push the throttle forward and you are ready to drive.

It was largely a man's world.

But give the Model-T all its due. It was suited to the times. It reflected a very reasoned and intelligent application of the then-available technology. It was mechanically simple and easy to maintain, riding high enough to operate on rutted rural roads, sufficiently lightweight to be pushed out when it was stuck, and inexpensive enough that it didn't require real affluence to own one. The whole enterprise was a success that had a marked and lasting impact on the nation.

You may have noticed that automobiles have changed since then. Now the electric self-starter has finessed this daunting process of manual starting and made it easier to drive than to harness a horse. Manual spark and choke controls have disappeared. The automatic transmission, by contributing to the emancipation of the less operationally versatile driver, further encouraged the expansion of the driving populace. And the computer chip has already started its near-revolution in the continuing automation of vehicle functions.

But other things have changed very little...

We still step on the accelerator to go, step on the brakes to stop, and turn the wheel to steer. More important, driving, while more comfortable and requiring less muscle, still demands the full-time attention of the driver and essentially the same skills. While the functions of the vehicle itself have been progressively improved and automated to require less and less attention from the driver, the functions of driving the car have changed very little. The demands on driver vigilance have not changed at all; in fact, with increasing traffic and higher speeds, they have increased.

We are now on the verge of revolutionizing the world of driving.

We have already started in a modest way. Automatic cruise control is now almost standard, and most of the driving public is thoroughly familiar with this automation of the job of holding constant speed.

Increasingly a new and different kind of cruise control is being offered purchasers of some of the more upscale new cars. This new cruise control will do more than just hold speed where the driver selects it. It will also decide if the car is about to "cruise" into the rear end of a slower car in front and slow one's vehicle so that doesn't happen. The driver doesn't need to do a thing.

The first models of this new "intelligent" or "adaptive" cruise control use the throttle and gears and perhaps just a little brake to slow the car. But maybe by now some are available with full, emergency braking capability. It's just a matter of time.

Aside from the enhanced safety, this advanced cruise control should make driving in traffic a much less demanding experience than the constant accelerator-brake jockeying to which many of us freeway drivers have become reconciled. In traffic the driver can set this intelligent cruise control to hold a constant distance from the car in front: if it slows, you slow; if it speeds up, you speed up; if it stops suddenly, you stop also— as suddenly as necessary to avoid a collision. Intelligent cruise control will decide, based on one's speed and a few other variables, what is a safe following distance and won't let your car get any closer than this distance.

Many—maybe most—freeway accidents are rear-enders, and most rear-enders appear to be the result of driver inattention: the kids were fighting in the rear seat, or the radio needed retuning, or the driver was admiring the scenery, or had momentarily dozed off. While intelligent cruise control will probably be no safer than an alert and prudent driver, very few drivers are alert and prudent all the time. The sensors in an intelligent cruise control system do not get distracted, get tired, or fall asleep. Intelligent cruise control will be far safer than the driver who is not at the moment alert and prudent.

There is no magic involved. Various sensors are added to the car to "see" the traffic ahead. These sensors—maybe a small radar or sensors much like those used in a camcorder—feed a little computer that interprets what the sensors see and signals the brakes and the throttle to slow the car or to speed up. Where a human driver uses eyes and a very small portion of his or her brain to interpret the traffic condition and tell the feet which pedal to push, the new intelligent cruise control uses this much less intelligent but much more focused "computer brain" which "sees" and interprets and tells the power brakes and power throttle what to do.

This all seems very sensible, and the basic technology needed has been around for a long time. We ask ourselves the question: "Why did it take so long to get around to doing it?"

We conjecture that there are three factors that played some role. First, the technical simplicity is a bit deceptive. Even though the technology is available and the idea is straightforward, it still requires some very sophisticated engineering and an awful lot of testing to put an operational unit on the street. The investment needed is large.

Second, the issue of liability is a real cloud hanging over that investment. With intelligent cruise control the auto manufacturers become much more vulnerable to the argument that the car caused the accident, not the driver, and past experience has shown that many juries are happy to assign blame to the deepest pockets. Product liability has to be a major concern, particularly in this country.

Third, the driving public has not been clamoring for it. Intelligent cruise control may be a bit like television: nobody knew they wanted it until they saw one. Invention is the mother of necessity.

Even given these headwinds, it looks like intelligent cruise control is finally happening.

Tip of the iceberg

Still, while the improved safety and added helpfulness that intelligent cruise control will provide is not trivial, it hardly constitutes a revolution in road transport. But it is the icebreaker, the first of the many steps that will, in our judgment, easily add up to a true revolution and watershed in transportation. As will be brought out in future chapters, the impact will spread well beyond the road system and its vehicles.

Hands-off freeway driving

But back to the highway. We conjecture that the next step, probably several more years downstream, will be the car that is capable of steering itself down the middle of the lane, with no help whatsoever from the driver. Such lane keeping systems have already been in tests for well over a decade now, both in this country and around the world, most notably in Japan and Germany. And systems that warn the driver that he is drifting off the road are already in operational use on large trucks.

This automated steering has been publicly demonstrated—along with many other automated features—in the Federal Highway Administration's National Automated Highway System Consortium (NAHSC) program [2]. Figure 2.1 shows such a vehicle in action during the NAHSC demonstrations.

At least in the beginning, this automated lane-holding system will work only on freeways and other limited access highways, not on ordinary city streets. The highways will require some minor modifications, like making sure the white lines are satisfactory for sensors on the vehicle to "see" and steer on and maybe to insert small magnets down the centers of the lanes for a magnetic steering sensor on the vehicle to follow. Figure 2.2 illustrates the latter scheme. This technique was developed by the PATH program of the Institute of Transportation Studies at the University of California at Berkeley and is the one guiding the car in Figure 2.1.

It's likely that the first operational systems offered for sale will incorporate several schemes to ensure safety in the event one fails. There are

Figure 2.1 "Look, Ma, No Hands." A demonstration of automated steering during the demonstration program of the NAHSC (Gerald Stone, California PATH Publications).

also other devices to ensure fail-safety, and they will probably all be used until experience tells us that some are unnecessary.

The driver becomes a passenger

With the combination of intelligent cruise control to handle the brake and the throttle and automated lanehold to steer in the lane, we now have hands-and-feet-off driving on any freeway or highway where the relatively inexpensive retrofits described have been made. The driver just drives to the lane he or she wants, turns on the automated system, and lets it "drive" until manual control is retaken. Tired drivers won't drift off the road or drive into the vehicle in front. Long trips become an opportunity to read a book or, after time verifies trustworthiness, even to nap.

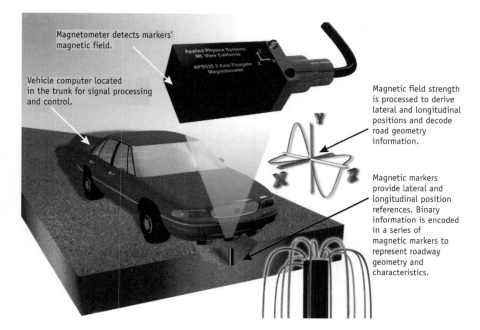

Magnetometer detects markers' magnetic field.

Vehicle computer located in the trunk for signal processing and control.

Magnetic field strength is processed to derive lateral and longitudinal positions and decode road geometry information.

Magnetic markers provide lateral and longitudinal position references. Binary information is encoded in a series of magnetic markers to represent roadway geometry and characteristics.

Figure 2.2 The primary features of magnetic lane guidance (Jay Sullivan, California PATH Publications).

All of these systems will be "failsoft," meaning that if they fail for any reason they will not cause some sudden maneuver but will warn the driver and provide ample time for him or her to take over the controls.

Even so, they may take some getting used to before most people will stop watching them like hawks, and there are a lot of people who won't want them no matter how well they prove to behave. And in the early years before volume production brings costs down, they will be expensive. But the safety improvement and the sheer convenience they offer will be hard to resist by anybody who can afford them, particularly those drivers who spend lots of time on freeways or long intercity trips.

These steps so far will have made a major contribution to safety, and they will have made driving a much more pleasant experience for many. But, except perhaps for some of the emergency stopping features, the systems will be confined to freeways and major highways where the lanes have been modified to accommodate them. So they will not have helped the nondriver. More has to come.

New, exclusive highways?

One idea that has been given very prominent exposure is the building of new, specially configured highways or new lanes for the exclusive use of automated vehicles. We think that idea will fade away.

The notion of not letting cars under automated operation use the existing highway and freeway system and constructing a completely new system just for the automated vehicles raises both a where-is-all-that-money-going-to-come-from problem as well as a terrible chicken-and-egg problem. We doubt that any auto manufacturer is going to make the very major investment of bringing these automated vehicles to market until there are many, many miles of roads where they can be used; why else would anyone buy them? And who is going to provide a lot of empty highway space until there are plenty of cars equipped to use that space? After you, Alphonse.

To demand that automated vehicles can't be used until there are special roads for them seems to us the modern equivalent of dedicating in 1900 all existing roads to horses and building new ones for the cars coming on to the scene.

(Actually, there was a scheme proposed about 1900 that provided separate, parallel roads for horses and horse drawn vehicles, for walkers, for electric streetcars, and for autos and trucks. Somehow the idea never caught on.)

Further, as we will discuss, we envision the day when vehicles will be capable of making fully automated trips rather than be confined to freeways. Requiring a completely separate road network for such vehicles entirely defeats this possibility.

So we think that these self-driven vehicles will operate in ordinary traffic with ordinary drivers from the very beginning. The first steps, already being taken, are reasonably modest; and each of the subsequent steps would profit from the experience of earlier ones. We believe that such a step-by-step evolution is by far the most sensible approach [3].

Continuing evolution

We won't try to drag you through every step we see in the continuing evolution of vehicle automation, but one can't be glossed over.

This is the migration of automated driving from just limited access highways to surface streets. This step will make it possible to have automation for the complete trip, not just cruising down a lane of the freeway or highway.

On freeways and limited access highways there are no intersections to worry about, just on and off ramps. So there the automation—except for the lane markers and little magnets in the road we mentioned—is all on-board the vehicle. But automatic operation in intersections is a bit beyond the automated system we have described so far and will almost surely require signals from the roadside to augment the vehicle's own "intelligence," just as drivers need traffic lights to keep intersections from being huge delays or junior-grade demolition derbies. On surface streets with intersections, the "smart car" will depend on electronic cooperation with the "smart street."

Today the traffic control system tells the driver when he can go or turn using traffic lights; the new traffic control will communicate directly to the vehicle itself and be "smart" enough to keep everybody out of trouble. Throw in pedestrians, and we begin to need a very smart system indeed. But like nearly every other complex system the world has ever seen, we will get there one step at a time, learning as we go, reaching capabilities that seemed totally out of reach at the beginning.

And this brings us very close to the fully automated trip, the car that can drive itself. Improved sensors to ensure better safety on streets where there might be children or pets, on-board navigation systems, and a few little frills to get us in and out of the garage, and we have the fully automated automobile. Now the car can take Dad to work or to catch his commuter train, come back home to take Susie to school, and stop by a car wash on the way back.

Designing the system: we emulate the human

We've already talked here and there about how these systems might be mechanized. Here we look at the issue from a broader and perhaps slightly different perspective [4].

We all know how we drive. We get most of the information we use to control our car through our eyes. We "see" the state of traffic, the position and movement of the other vehicles around us. We get lots of other visual cues: lane markers, traffic lights, signs, and rain on the windshield. Now this "seeing" is really two functions: the eye senses the world out there, and the brain interprets the data from the eye into the scene we "see."

We sometimes get relevant information through our sense of motion and sometimes through our ears, but with cars more and more insulated from sound on the outside and with competition from radios, car phones, and conversation on the inside, this latter source is a sometime thing.

Using all this sensed information in combination with our own "rules and guidelines" for driving, we—our brains—reach decisions on what actions are desired to control our own vehicle: do nothing, turn left, turn right, brake or speed up, and occasionally call for divine intervention. We then take the indicated action, operating through the vehicle's control system.

We find it impossible to think of any artificial system that doesn't basically emulate these functions of sensing, interpretation, decision, and action. How each of them is carried out, however, is another matter entirely—there are lots of possibilities. We will try to describe the technical landscape as we see it, focusing on the sensing and computation aspects of the problem.

As we noted, our primary sensor (for driving) is the eye. We know how to build sensors that can do essentially everything the eye can; we use them in camcorders and digital cameras of all types. But in the realm of artificial systems we have additional options that the human eye can't provide. First, unlike the eye, we can build sensors that work at wavelengths other than those of visible light. For example, we have the potential of using infrared wavelengths that can "see" through fog (they are a little expensive today, but the prospect is there).

Second, we also have more options for so-called "active" sensing over a wider choice of wavelengths. In "active" sensing we transmit electromagnetic waves to illuminate the object and then get information from the waves that bounce back. This is, of course, what headlights do, but headlights and eyes can't measure distance to the object ahead. At the lower frequencies used in radar or in laser ranging devices we can measure the distance to another vehicle with fairly high precision.

Third, by using multiple sensors we can have an "eye" wherever we want them, looking in all directions simultaneously.

So in a technical sense we are really not bad off in the sensing function; in fact, in this function we have the potential of being substantially better than the human. The Defense Department has spent billions over the last 20 or so years developing and improving sensor systems for military use, and this technology is daily finding its way into civilian applications—and on to vehicles of all kinds.

As we already implied, the eye is analogous to the sensors—the antennas—on our radios, radar, and TVs; it's just tuned to a different wavelength. These antennas all by themselves are pretty dumb; it takes the rest of the system to interpret the signals they pick up into the sound we hear or the distance to the car ahead or the picture we see on the screen. So the eye-antenna is not much smarter than the radar or the TV antenna. But the rest of the eye-antenna system—the brain—has been practicing for millions of years at interpreting its signals and has gotten pretty good at it. It's this interpretation that lets us "see" the whole scene and lets other parts of the brain make our decisions as to how to drive the car. But when we substitute artificial eyes, like a digital camera type sensor or radar, we also have to substitute some kind of computer brain for the human brain. That is a tougher act to follow.

Fortunately we don't have to approximate the whole brain, just the small part that interprets what the eyes "see" and the part that decides the desired action to take. In the early evolution of our automated system, both of these jobs are well within the current state of the technology. For example, the three primary pieces of information that intelligent cruise control needs are (1) distance and rate of change of distance to the car ahead, which we can get with radar that has no trouble translating its signals into these values; (2) speed, which can come from our speedometer; and (3) road surface condition for braking, which can be deduced several ways. The decision function is based on calculating the separation distance necessary to ensure no collision under whatever ground rules have been specified and then how much deceleration or acceleration is needed to maintain this distance. Not simple, but straightforward; probably much easier to do than developing Windows 95.

As we continue to add more complex automation, we will also require more of our computer brain. But, like always, we will add complexity one step at a time, each step from a little higher base of knowledge and experience. Just a glance at our progress in computation in the last decade makes it hard to believe that this system is beyond our reach.

The end of driving?

No, this automation option won't be the end of driving. There is no reason to preclude the manual option; in fact, as we will point out, having the automation should make it more attractive.

Fully automated trips can only be made in areas where the advanced and integrated traffic control system is in place to support this kind of operation. It's probable that there will always be large rural areas that cannot support such full automation; Mom, Dad, and Susie will have to manually drive their car for their weekend picnics.

The nondriver isn't anymore

But while the fully automated vehicle is not the end of driving for drivers, it might well be the beginning of driving for nondrivers, those folks who now do not drive because they are too young, or too old, or too fearful, or just incapable. The computer brain allows the unskilled to drive safely: if all is going well, the automated system only sits and watches; but if the driver does something unsafe, it intercedes. This safety-watchdog function should make it possible for almost anyone to safely drive a properly equipped car.

With the automated car, we are taking a very big step toward good, convenient mobility for all.

Our new automated vehicles cannot only help people drive, it can teach them how. Each year our teenagers learn to drive. Why not let the car teach? If the apprentice driver tries to do something unsafe or unlawful, the controls will thwart; if something imprudent is tried, a voice will give a bit of a lecture. The teaching car would make a nice niche market.

Neighborhood design and home design can change. It will no longer be necessary to build an extra bedroom for the car. Here we have a car that self-drives, so it can clearly self-park. Now designers can think even more about communities where parking is at the edge and out of sight. The car can drop you at your own door and go park itself, especially nice on those cold, dark, rainy nights. Ready to go again? Call up the car to the front door.

The automated car offers more than just automation

The automated car will require so-called drive-by-wire controls; that is, the brakes, throttle, and steering are actuated by powered devices in response to electrical signals, not driver muscle. This isn't far-fetched; drive-by-wire is already appearing in some cars. Some European Airbus aircraft and the Boeing 777 operate routinely with fly-by-wire controls.

But once the car can be driven by electrical signals, driver muscle is no longer needed. Now we don't need a steering wheel, which started as a way to let driver muscle turn the wheels, and we don't need pedals on the floor. The way the driver controls the car is wide open for new approaches. We've heard of a Mercedes concept car that is driven by a small joystick: move it forward to go, back to stop, and sideways to turn. Car design can enter a new era.

Platooning

Somewhere early in this evolution, we'll see a new capability added called "platooning." This is the capability for properly equipped vehicles to automatically form "trains" but with electronic connections between them instead of the mechanical ones used on railroads. This would work by having each vehicle automatically communicate with adjacent vehicles so that they could coordinate their controls, which would allow them to operate safely very close together.

The objective of such entraining or "platooning" is to increase the effective capacity of the highway by increasing the density of vehicles [5]. When a lane is getting crowded, adjacent vehicles will automatically move together to effectively form a single unit and thus make room for more vehicles to enter the lane. When a vehicle needs to exit the freeway, the platoon would pull apart and let the individual vehicle leave the platoon.

When a substantial proportion of vehicles are properly equipped for platooning, we can at least double the effective flow capacity of a freeway lane. Some researchers are even more optimistic about the capacity gains that will be achieved, but even doubling the capacity of our freeways without adding more concrete is an extremely big carrot!

There are very legitimate questions about anxiety levels riding in a car that's driving itself 2 ft or even 20 ft from the next car at 70 mph. Lots of people have tried it (maybe a bit under 70 mph) as part of the Federal Highway's automated highway demonstration program, where this platooning capability played a prominent role. The general judgment seemed to be that most people could get used to it. Time will tell.

It's not just for moving people

All this automation and design change constitutes not only a radical change in the way we think of the car, it could revolutionize the way we think about urban delivery and goods movements in general. We will conjecture a bit about some possible implications, and we invite the reader to conjecture with us. After all, it will be a world new to us all.

The pickup and delivery vehicles in our new world don't need drivers. So they also don't need padded or protected compartments; they don't need steering wheels, dashboards, windshields, brake pedals, accelerators, gear shifts, running boards, doors, heaters, air conditioning, or windows that roll down—or even windows at all. We are so used to these features being fundamental to any vehicle design that it's hard to imagine what vehicles might look like when they are no longer necessary.

Because the automation gear—the computer brain and the sensors and the power actuators—don't weigh much in comparison, these new delivery "bugs" should be very much smaller than a conventional vehicle for a given load capacity. They will, therefore, need much less power than the urban trucks of today. Thus they should be extremely parsimonious with their use of energy and have very, very small adverse impact on air quality.

They should be cheaper than today's trucks because the structure should be lighter and simpler, and in high volume production the automation gear should have come down in price just as we have observed already in TVs, VCRs, computers, and cell phones. It should be relatively inexpensive for even a moderately small commercial enterprise to maintain a range of vehicles specifically tailored and sized to specific jobs.

This same line of reasoning should apply to intercity movements as well. While we get a little squeamish about the idea of 18-wheelers roaring over the highways with no drivers, we have less trouble with smaller "trucks" specifically designed for this job.

An important part of the reason large trucks are less expensive than small trucks on a cost-per-ton-mile basis (or trains over trucks or big ships over trains) is because labor and other fixed costs are spread over more "tons." Eliminating labor costs thus makes the small "truck" more cost competitive with the large truck. The net result is that the cost penalty for moving things in small batches between cities will be substantially decreased.

The lower cost of small batch transport permits more frequent service than if the higher cost of transport dictates only larger but less frequent shipments. Small, frequent shipments reduce the size of the temporary inventories awaiting movement. It further enhances "just-in-time" opportunities.

Small batch transportation accommodates the nearly universal trends toward more personalized production and delivery—the computer to your specifications, for example. The automated delivery vehicle fits beautifully into this trend.

An easier and cheaper small batch capability will help widen the already existing urban delivery market. We can envision driverless vehicles filling an increasing niche, especially for those goods that can be automatically loaded and unloaded.

Thus the biggest impact may not be just the drop in costs of small batch movement relative to large batch movement but the secondary impacts that follow from doing old things better and perhaps doing entirely new things. This is a theme you will hear repeated over and over throughout this book.

There will obviously be problems to overcome. For example, driverless delivery of retail orders of groceries to the home will not be practical until there is an alternative available for the driver's role in taking the delivery into the home; perhaps some kind of a secure, small package receiver box accessible from the curb, much like the curbside mailbox or newspaper receptacle. This sounds easier than it will be: we have to worry about things like the ice cream melting. But we'll bet that some entrepreneur figures out a way to cope with such difficulties.

Brave new world!

There is no shortage today of predictions telling us how we will live in this technologically new world. With tongue slightly in cheek, we offer a composite summary of some of these various views that are relevant to our subject.

Everybody will do all their work from their very own home office, do all their shopping on the Internet and have their purchases delivered to their very own home, get all their entertainment through electromagnetic waves to their very own media room, and invest all their money in the thriving Couch Potato Mutual Fund. Their cars will just sit wherever they parked themselves and gather dust.

You say you don't think it will work out quite this way? We confess that we don't either, even though there may be some tiny grains of validity buried in there.

What we do predict is that even with our truly amazing progress in communication and transportation, people will continue to be people. The change will be in the richness of their menu of choices and options as to how to live their lives. We are not brave enough to predict the net impact of this remarkable new world on our general societal behavior, but we are willing to assert that, contrary to our little fast-forward scenario, we will not become immobile.

Now in this chapter we have introduced the automated highway vehicle. We close the chapter by summarizing the specific impacts we believe this watershed step will have on our lives.

First, they will make driving easier and safer for everyone.

Second, they can open up a whole new world of personal mobility to nondrivers.

Third, they will significantly increase the effective capacity of our existing road and freeway system.

Fourth, they will open the door even further to a completely new category of vehicles based on designs that are not constrained by the need to have provisions for manual control or even human occupants but are sharply tailored to the particular niche application desired.

Last, the fairly dramatic impact on the cost of small batch movements will precipitate a rethinking of manufacturing and maybe even social processes, potentially leading to further improvements in overall productivity, efficiency, and quality of life.

The car that can drive itself? Ridiculous? Just you wait.

Endnote.   We are far from the first with such ideas. In 1940, Mr. Norman Bel Geddes in his book *Magic Motorways* laid out his vision of an interstate highway system that would support high-speed, fully automated vehicles [6]. He saw automation as the path to much improved safety and efficiency. His were the ideas for the highway system of the future depicted in General Motors' very popular Futurama exhibit at the 1939 World's Fair in New York. Thus he not only beat us here, but he was already ahead of our later chapter on superspeed highways.

References

[1] MacMannus, Theodore F., and N. Beasley, *Men, Money, and Motors*, New York: Harper & Brothers, 1929.

[2] Ioannou, Petros, "AHS Activities," *Automated Highway Systems*, Petros Ioannou (ed.), New York: Plenum Press, 1997, pp. 1–10.

[3] Yim, Youngbin, et al., "Integration of Automated Highway Systems into Existing California Freeways," *Automated Highway Systems*, Petros Ioannou (ed.), New York: Plenum Press, 1997, pp. 29–48.

[4] Ward, Jerry D., "Step-by-Step to an Automated Highway System—and Beyond," *Automated Highway Systems*, Petros Ioannou (ed.), New York: Plenum Press, 1997, pp. 73–91.

[5] Shladover, Steven E., "Reasons for Operating AHS Vehicles in Platoons," *Automated Highway Systems*, Petros Ioannou (ed.), New York: Plenum Press, 1997, p. 11.

[6] Geddes, N. B., *Magic Motorways*, New York: Random House, 1940.

3

Congestion: The Devil We Know

The automobile has been a big success, but there is a price.

A dispassionate observer—say a man from Mars—might well agree that the fully automated vehicle story suggests a truly tantalizing future. But we suspect he might drag us back to more immediate problems.

"Look, now is now. You still have a decade or two to wait for the big payoffs from automation. In the meantime you have to live with today's realities. The real devil you've got out there today is urban congestion."

We will devote the next three chapters to this subject. Here we assess its place in our lives. In Chapter 4 we summarize what's being done now to alleviate it; most of these steps are part of the intelligent transportation system program. In Chapter 5 we offer some approaches for going still farther toward driving this devil from our lives.

Ultimate gridlock

Let's first dispense with one myth about the congestion devil we are to drive out.

Many people fear that this devil is going to take us to the hell of ultimate gridlock. They seem to envision that someday just one more car will drive onto the freeway, all traffic will stop, and everybody will get out of their cars and walk home and stare at the walls. The city will cease to function until the whole mess can be paved over and we start again. Congestion really is a devil, but this ultimate gridlock is not going to happen.

Think about it. Long before this gridlock, folks will change what they do. When people have easy-to-exercise choices and they have a hard time getting somewhere, they find somewhere else to go. We have seen that in our cities. In the early suburbs, suburban dwellers continued downtown shopping and church and theater going; but when the traffic got tough, the tough went elsewhere, so to speak. They began to shop, pray, and play in the suburbs. Change is easy when the traveler has lots of destination choices.

But many folk do not find changing jobs very easy, they must cope in sometimes trying ways. When the 30-min commute becomes an hour, people start to sell their extraurban homes and move closer to their jobs or move away entirely to a smaller city. Some employers move out nearer their employees or set up branch facilities, and people change the scheduling of travel. And we expand the road and freeway system to help people cope.

Touting the congestion devil and ultimate gridlock supports those who want solutions and those who search for solutions. That's understandable. The shopkeeper concerned about customers' access asks the city council for relief. Congestion provides a continuing pressure for more money to our highway planners and builders. It is the centerpiece of the argument for more transit. It's meat for environmental groups, some of whom see it as a reason for policies to decrease auto use, others who push for a cure because congestion results in more fuel consumption and exhaust emissions. Traffic engineers and traffic policy buffs know that eliminating traffic jams increases highway capacity and press for treating the congestion disease.

There is no question that there is a devil out there. Congestion bothers just about everyone, and when it gets bad enough people will take actions that limit its further growth or their exposure to it. People and businesses may not like to do these things and employers and merchants and city fathers may not like to see them done, but congestion is itself a motivator that finally limits its ultimate severity.

But the fact that congestion is ultimately self-limiting is a very small comfort: it still leaves us with an onerous problem.

Is congestion increasing?

There is nothing new about congestion. Congestion in the emerging modern world caused by workers arriving on time for jobs in factories began with the proliferation of clocks and watches in the eighteenth century. That was when work began to be organized by time rather than by tasks. American school children learn that the push of congestion played a role in settling the American West; they learn that Daniel Boone left Kentucky for Illinois when he felt that he had too many neighbors. James Winter tells us rich tales about work, play, and lifestyles on Victorian London's teeming streets [1]. Earlier, the City of London had to deal with neighbors' complaints about unruly crowds in the vicinity of Shakespeare's Globe Theater—the NIMBY (not in my backyard) complaint is not a new one.

Deep down, congestion is a symptom of success. Unpopular places are seldom congested. And congestion due to place-popularity is compounded if it is also accompanied by time-popularity, as when everyone wants to use the same freeway at the same time of day—the work schedule situation we noted earlier. Its ramifications also turn on how things are perceived and folks reactions. It's a messy subject.

Many have the perception that congestion has been getting worse, and in some general sense that has to be true simply as a result of population increases—there are more folks to cause congestion and more to be impacted by congestion. To move beyond "some general sense," let's ask when, where, and how the severity and extent of congestion has been getting worse. It's a relatively complex story because so many variables enter into the picture.

Our first clue is the relationship shown in Figure 3.1, where the growth of urban vehicle-miles driven is compared to the miles of lanes available to accommodate this driving [2]. We show miles of lanes rather than miles of road to account for the fact that considerable capacity increase comes from the addition of new lanes to existing roads and freeways.

Figure 3.1 speaks for itself, but we can't resist adding a word or two. Vehicle-miles per day can increase either because of higher average speeds or because there are more cars on the road. Since vehicle-miles are growing more than twice as fast as lane-miles, we can't begin to account for the increase through speed alone. There are more vehicles per mile of lane; that is, the average density of vehicles on the roads and freeways is increasing. We will show in Chapter 5 that vehicle density is a key indicator of the likelihood of congestion.

Other data confirm this deduction. Hours of delay due to congestion have been compiled by the Texas Transportation Institute for our 50 largest cities, and these data are included in *National Transportation Statistics 1997* [3]. Data show that the overall average delay from congestion has risen from about 3.2 min per licensed driver per day in 1982 to roughly 5.5 min in 1993, an annual rate of growth of 5% [4].

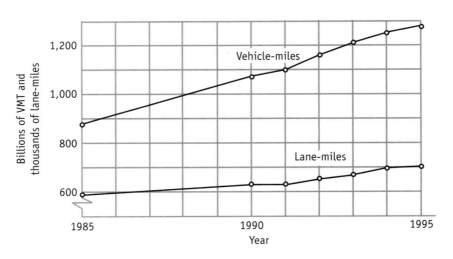

Figure 3.1   Urban vehicle-miles driven and lane-miles available, 1985–1995. [2]

If that 5% growth rate has continued, then the congestion delay per licensed driver is almost 8 min in 2000 averaged over the 50 cities. We'd rather not have to put up with any, but 8 min per day doesn't sound so terrible. That is, until we remember that everybody with a drivers license doesn't drive every day and many of those who do are not in the part of urban network that is overcrowded, so many actual drivers are getting a good bit more than their share.

But, contrary to most of our impressions, it is not the commuter. In a 1994 paper Professors Peter Gordon and Harry W. Richardson at the University of Southern California tell us in detail about travel by individuals [5]. They have examined travel schedules, times, and speeds in metropolitan areas as measured by travel surveys and paid attention to the growth of travel in the suburbs versus that in central cities. After looking at data for cities of various sizes, they conclude that the adverse impact of congestion on individuals in their journey to work is, on average, not increasing appreciably.

In the 1980s, for example, the average one-way commuting trip, a trip in the ball park of 20 min to 25 min for most cities, increased by about 40 sec. In the same period, urban vehicle-miles of travel (VMT) increased by about 50%.

A paradox?

How can this be—lots more travel, more congestion overall, but hardly any difference in average travel times for the commute trip? Gordon and Richardson and others point to increased travel in the suburbs where there is less travel hassle. Folk adapt by moving away and changing travel patterns. That's the main bottom line. But as Alan Pisarski's interpretation of the 1990 National Personal Travel Survey tells us, it is set within the context of changes in the size and composition of the work force, shifts in the modes used for travel, and other things [6].

So more and more commutes are largely within or between these suburbs, where trips, on average, are both shorter and faster. But those drivers that have to commute into the central business district spend roughly twice the time as in the suburban commutes. Even with congestion over the whole urban network increasing, the proportion occurring where it is most consistently onerous—the suburb-to-CBD commute—is decreasing.

The authors live, respectively, in suburbs of San Francisco and San Diego, and to us congestion seems to have gotten a lot worse than 40 sec on a 20-min trip. We sometimes see our freeways during rush hour looking like parking lots and wonder. We don't doubt that the Gordon–Richardson and other data are valid. The apparent paradox simply lies in the fact that averages can hide wide variations from place to place.

We think of the old story of the man drowning in a stream where the water is, on average, only a foot deep. It may be only 6-in deep in many places, but those 7-ft-deep holes are downright dangerous. Analogously, congestion may not be bad everywhere, but in places it can be a doozy.

Think about patterns and dynamics of change in a single suburban area, where there can be substantial changes in both trip patterns and congestion levels as it evolves over time. We'll sketch out this progression here, and again in Chapter 7.

A short story about the suburbs

Suburbs are a phenomenon that began on a very small scale well over a hundred years ago, but the serious expansion of auto-based suburbs, held back by the depression in the 1930s, did not start until after World War II. Then we had a wave of new family formations, the widespread availability of automobiles for transport, savings accumulated during the war years, rising incomes, and government programs favorable to new home construction and ownership. This expansion of people and homes into new suburbs was naturally accompanied by the construction of more freeways, arterials, and local access streets.

Let's go back to 1948. You and your family has just moved out to the new housing development in the country outside the city. It's great! Your home has a backyard for the dog, for the barbecue, and for the baby when she's old enough to go outside alone. It is kind of a drag to have to drive so much farther to your law firm downtown, but the new road is good and not too crowded.

It's fine with your husband, too. There's that new shopping mall being built just two miles away, so when it's finished he won't have to drive all the way into town when he needs more laundry detergent or something.

Spin the hands of the clock. It's 1982. You are going to retire next year. You hate to drive the freeway because the new suburbs being developed farther out are making the traffic worse.

Years ago your law firm moved to the new office buildings just a mile away, so that old commute back into the city is no longer necessary; you only have to use the freeway in order to visit your daughter and her family who live 9 miles further out.

Your new husband Franklin does have to use the freeway; he works at the newspaper printing facility that moved out of the city twenty years ago to be near the new suburb—it was cheap land then. The commute is 8 miles on the freeway and the total trip takes him about 16 minutes. Not bad, but he complains it is getting worse.

It's three years later. You've found that if you visit your daughter in the early afternoon, you can avoid most of the congestion on the freeway. Your husband is due to retire in three years, which is lucky because there are so many more people using the freeway that the average speed is closer to 30 mph than the old 60, and his commute is now over 25 minutes and apparently getting worse still.

Six months later Franklin knows they are planning to widen the freeway, but the thought of driving it for another two years during the construction is more than he can stand. He has decided to retire two years early. The two of you have already made the down payment on that half-acre lot just 5 miles beyond your daughter's home.

Let's take a top down look at what we've just described. People moved to the suburbs. Had nothing else changed, this movement of large numbers of people out of the old city would have resulted in a marked increase in average trip distances and vehicle miles of travel since workers would have to commute to their jobs in the city and shoppers to the stores. The average trip would be much longer than in the old days when people both worked and lived in the original city.

But, as we noted, other things did change. Retailers followed the people to the suburbs; the shopping mall is one manifestation of this trend. Second, taking advantage of cheaper land, some manufacturers also moved to the suburbs, automatically moving the jobs with the factory. Third, services such as banks moved out where the people lived. So jobs and services followed people to the suburbs, shortening commutes and most other trips.

So less and less of the increased traffic was in and out of the city center and more and more within a suburb or suburb to suburb. The result was that average travel times actually changed very little, and for most of this time suburban road and freeway expansion was largely able to keep up with the growing number of trips deriving from the growth in population.

That's part of the dynamic—travel reorienting from the old city to the suburbs. The other part is that as suburbs build-out, new ones are created. So the impact of congestion has been tempered as one suburb after another has been developed and roads and freeways progressively upgraded.

As would be expected, the situation differs from city to city. Differences are seen in the 1997 *National Transportation Statistics*, where the data are presented city by city [3]. The report provides data on the percentage increase in total delay in person-hours in the average commuting trip between 1982 and 1993 in 50 of our larger urban areas. As would be expected, the rate of growth varied widely among the various areas, with the smaller ones generally showing the most rapid growth. The delay in the worst 18 cities in 1982 had grown by 1993 by 86%; in the 32 smaller cities the delay grew by 146%. The combined delay in these 32 smaller cites made up only 18% of the total.

What congestion is costing us

Researchers at the Texas A&M University, whom we mentioned before, have collected and interpreted data for cities since the 1980s [7]. Congestion is measured in hours of delay and the cost of delay. Congestion delay is assigned a cost of about $10.00 per person-hour, and total congestion costs include excess fuel cost due to stop-and-go driving. About one-half of delay reoccurs from day to day and about one-half—more in many areas—is from incidents such as accidents [8].

We see the findings from the Texas A&M work and similar work on the TV and in the newspapers, "Congestion cost residents of East Cupcake millions of dollars last year…." What we see are big numbers, big dollars; in 1993 these costs typically ranged between $150 and $800 per capita per year [3]. Total costs go up and up over the years, largely because the population is growing and so has VMT; more people are exposed to the congestion, even though the average severity on the average trip may have changed very little. Population has been increasing by about 25 million persons per decade. That adds about 10 million households. With an average of two vehicles of all types per household, about 20 million more vehicles are using roads each decade.

Numbers such as these suit the "just give me the facts" attitudes of many, and the authors usually have that "facts" orientation. Yet the "facts" don't seem to fully describe the costs imposed by the congestion devil.

It's the cost of coping, of opportunities foregone, that matters. With congestion, getting here to there takes more time. Stephen Fox tells us about Walter P. Chrysler's interview in a 1927 issue of *Colliers* magazine in which he used the word speed 16 times and said, "By speed I don't mean breakneck travel for the sake of thrill, but quickness in getting somewhere to do something useful quickly" [9]. Data on annual costs for a mythical average individual are one thing. To say that congestion puts limits on "doing something useful" (and thus the surprising ways Americans use time [10]) describes costs differently.

There is another point about cost. Congestion puts the path of urban development on a build-new and move away from the congested-old basis. The costs are real and we should consider them, along with the penalties incurred by folk who could not or did not want to play the live-in-the-suburbs story we sketched.

The overall lesson is that our inability to better manage congestion limits our choices. The corollary is that improved transportation improves our lives by increasing choices, and we will make much of that point as the discussion goes along.

Just part of the story

When a freeway or a road is really congested, the ability of that freeway or road to carry traffic—its flow capacity—is reduced. It has lots of vehicles on it, but they aren't going anywhere—at least not very fast—so the actual flow is lower, sometimes much lower, than if that road or freeway had fewer vehicles moving faster.

We taxpayers have paid a lot to build those freeways, and by letting them congest we are just throwing away part of that investment. It's as if we have spent our money putting in four lanes in a freeway and then closed down one or two of them. Congestion is sheer waste, and it takes some very contorted reasoning to find any redeeming features.

The loss in flow capacity from congestion depends, not surprisingly, on its severity. We were told some years back that in Los Angeles the "average" congestion cuts flow from the normal uncongested maximum of 2,000 to 2,500 vehicles per hour per lane to something like 1,400 to 1,600. This is roughly equivalent to closing one-lane on a three-lane freeway. Put the other way, eliminating congestion could effectively add another lane to a two-lane freeway, and for a lot less money than it would take to actually build that extra lane.

In really bad congestion, where speed oscillates between a dead stop and maybe 10 mph to 15 mph, the penalty is even worse. Flow under such conditions is cut even further, to more like a 60% or 70% loss. (As an aside, we should note that to our knowledge, actual data on the loss of capacity to congestion is very sparse, and our assertions are largely based on knowledge of the relationships among flow, speed, and vehicle density on a freeway lane. We think it would be illuminating if systematic measurements of this capacity-loss phenomena were part of normal traffic statistics.)

Economists often talk about the problem of "how to divide the pie." Our "pie"—the use of the freeway—has an unfortunate characteristic: if too many people want to share the pie, the pie shrinks. We can build three- or four-lane freeways, but because we let them congest, we only get the capacity that two or three uncongested lanes could have supplied.

So in addition to the price we're paying in lost time, frayed nerves, and increased pollutants, we are paying an additional price in lost freeway and road capacity. Because we let congestion persist, we are not getting our money's worth on our road-building investments.

And we hate to think we have to adjust our lives around this chronic urban phenomenon.

We can do better

We are not powerless in the face of this phenomenon; we believe we can go much farther than we have in alleviating this problem. We are not without options. But there's no magic bullet, and some of the measures carry costs in themselves.

And that is the primary point of our story. There is a lot to tell, and we have chosen to tell it in pieces. In the next chapter we describe the principal actions that are being taken today. It's fairly clear that it is not enough; the state of congestion in our larger cities tells us that. So we follow with some specific suggestions for doing better.

References

[1] Winter, J. H., *London's Teeming Streets: 1830–1914,* London: Routledge, 1993.

[2] Bureau of Transportation Statistics, *National Transportation Statistics 1997,* Washington, D.C.: U.S. Department of Transportation, p. 25.

[3] Bureau of Transportation Statistics, *National Transportation Statistics 1997,* Washington, D.C.: U.S. Department of Transportation, p. 57.

[4] Calculated from [2] and licensed driver data from Federal Highway Administration, *1993 Highway Statistics,* Washington, DC: U.S. Department of Transportation, 1994, pp. II–8.

[5] Gordon, P., and H. W. Richardson, "Congestion Trends in Metropolitan Areas," *Curbing Gridlock: Peak Period Fees to Relieve Traffic Congestion,* Vol. 2, Washington, DC: National Academy Press, 1994, pp. 1–21.

[6] Pisarski, A., *Nationwide Personal Transportation Survey: Travel Behavior Issues in the 90s,* Washington, DC: Federal Highway Administration, U.S. Department of Transportation.

[7] Hanks, W. J., Jr., and T. J. Lomax, *1989 Roadway Congestions and Trends,* Research Report 111-4, College Station, TX: Texas Transportation Institution, 1997.

[8] Bureau of Transportation Statistics, *National Transportation Statistics 1997,* Washington, DC: U.S. Department of Transportation, p. 58.

[9] Fox, S., "I Like to Build Things," *American Heritage of Invention and Technology,* Vol 15, No. 1, 1999, p. 23.

[10] Robinson, J. P., and G. Godbey, *Time for Life: The Surprising Ways Americans Use Their Time,* University Park. PA: Pennsylvania State University Press, 1997.

4

Congestion: What Are We Doing about It?

Sisyphus would have felt right at home trying to control traffic.

The automobile and the truck have become powerful examples of invention becoming the mother of necessity. They are victims of their own success, and one cost of this success is the traffic congestion we discussed in the last chapter.

What causes congestion? It's simple: more vehicles wanting to use the freeways and streets than the freeways and streets have the capacity to accommodate.

The road and freeway "capacity" about which we are talking here is not measured by how many cars we can get on the streets and freeways; that's how we would characterize a parking lot. Here we refer to flow capacity: how many vehicles can drive past a given point in an hour, much as we would talk about the flow capacity of a pipe in terms of the gallons of water that could flow out the faucet in an hour.

Just to give some feeling for the situation, one lane of an uncongested freeway can accommodate up to somewhere around 2,000 to perhaps 2,500 vehicles per hour. It can be thought of as telling us that roughly one vehicle drives past every 1.5 sec to 1.8 sec. Traffic engineers call this interval between vehicles the "headway."

On a typical arterial lane, where the traffic has to cope with intersections and traffic lights, maximum flow per lane drops to roughly half these figures [1].

These numbers represent about the best that can be done with today's cars and drivers. Obviously they are highly variable with lots of factors: street and freeway design, weather, local habits, and, of course, the degree of congestion. Throw in an accident, and all bets are off.

The broad prescription for congestion relief

Since congestion is the consequence of more car and truck drivers wanting to use the freeways and streets than the freeways and streets have the capacity to accommodate, the prescription is almost a no-brainer:

> To increase road and freeway capacity as much as we can within the many constraints of real life, and then control, regulate, or otherwise manage to keep usage levels within that capacity.

Remember this prescription; it is very important. As we describe what we are doing today, we will figuratively hold these actions up to the light to see how they fit this prescription of what is needed.

We can tell you now what we'll find out: nearly everything we do is aimed at the "increase capacity" part and not nearly enough on the "manage usage levels" part. In spite of our promoting public transportation and of the great publicity given the carpooling and various other schemes, we are failing nearly everywhere to adequately perform this part of the job, to control street and freeway usage sufficiently to get full benefit of its inherent capacity.

No matter how cleverly we blink the traffic lights or how smoothly the uncongested freeway flows, if we let the number of vehicles on the facility exceed the maximum number of vehicles that the facility can handle, we risk congestion. And once congestion starts we're in the soup; it takes a very long time to decongest. Recall the point made in the last chapter: letting a street or freeway congest cuts their flow capacity—our "pie" shrinks. This fact puts a real premium on preventing congestion from ever getting started.

Now we know the dictum to "control, regulate, or otherwise manage" usage levels has an ominous, big brother–like sound. And where congestion is really severe, where lots of vehicles are just waiting in the wings for it to let up just a little, it deserves a little of that reaction. But don't throw out the baby just yet; some of the options are relatively painless. We'll go through the whole gamut in the next chapter; now we'll describe the things that are currently being done to make life easier for the driver.

Traffic control and the intelligent transportation system program

In recent years there have been major efforts throughout the developed world to apply our new technological capabilities in computing, sensing, and communication to improve road transport. The effort in the United States has been dubbed the intelligent transportation system program, and this program has sparked an intensified focus on upgrading traffic systems already in place and on adding new capabilities of many varieties. It is a very broad and flexible program, and we only describe here those elements that we deem immediately relevant to our subject [2–4].

Traffic management systems

The most familiar technique for controlling traffic flow is traffic lights. These not only prevent—or at least reduce—accidents at intersections, their timing is set to try to maximize the number of vehicles that can get through the intersection on each full cycle and, thus, reduce the number of stops and waits.

A decade or so ago nearly all traffic lights had timing cycles that were preset and operated by the clock no matter what the actual traffic conditions. These settings were usually based on past observations of the average nature of traffic. If traffic is not "average," then the cycle is not optimum.

More and more in recent years sensors of various kinds—loops in the road, overhead TV-like viewers—have been installed to provide these systems with real-time information about the actual state of traffic. With such sensors, a traffic light at an intersection now can "know" how many vehicles are coming from each direction and how many are waiting in lines for a chance to go, so its timing sequence can be continuously and automatically varied to optimize the flow through the intersection for the specific traffic situation. Such systems are referred to as "adaptive."

Coordinating lights at successive intersections to provide a "green wave" for the dominant flow is an old trick, but that too can be made to work better if the timing can be based on real-time conditions.

Our traffic systems are very good at increasing the flow of vehicles through intersections and along arterials, but there is no attempt to use them to regulate the number of vehicles that try to get through an intersection or try to drive the arterials. Very few of our traffic control systems today are designed to exercise any control over levels of traffic.

Ramp metering is a special kind of traffic light designed to regulate the flow of vehicles onto a freeway: one car—or sometimes two at a time—every X sec. This not only regulates the rate at which vehicles can enter the freeway, it also smoothes their entrance. Experience has shown that smoothing the flow into the moving traffic not only cuts the number of accidents at entry but tends to smooth the flow on the freeway itself. The net result is both better safety and an increase in freeway capacity.

A few years ago the Minnesota Department of Transportation noted that the use of ramp metering in the Minneapolis–St. Paul metro area was the intelligent transportation system element that gave the single greatest boost to freeway capacity and safety [5].

But ramp metering systems are not generally set up to exercise precise usage control over the whole freeway. We have the technology to adapt them to do so, to keep the numbers of vehicles within the capacity of that freeway to accommodate. As we will develop in the next chapter, such advanced ramp metering becomes a key element to congestion prevention and alleviation on freeways.

On-the-roll toll collection

Systems are being put in place all over the country—and the world—to collect tolls without requiring the vehicles to stop or even slow down. Some of these automated toll collection systems work using short-distance radio signals to automatically deduct the toll from a prepaid card carried in the vehicle; some use bar codes or electronic identification devices on each vehicle so that the owner can be sent a bill at the end of the month.

These collect-on-the-roll systems are a huge help in decreasing the hassle and time lost at toll collection points and, therefore, contribute very significantly to increasing the effective capacity of toll roads. They also lower the number of rear-enders at toll collection stations. In the next chapter we will discuss an additional role for these devices [6, 7].

Driver and traveler information systems

How many times have you read in the news about this marvelous new system that will tell all drivers where the congestion is so that they can all go around it? "Lots" is the answer if you've been paying attention; this is probably the most widely advertised feature of the whole intelligent transportation system program.

We are dubious. Unfortunately, the driver often has no real choice of an alternative route even with the best information. Except in a few locations, the parallel surface streets just don't have the capacity to carry freeway-level traffic. And it's very difficult to choose another route when you're in the middle of the Golden Gate Bridge.

But information can be helpful even if it only tells the driver why he or she is sitting in a jam and how long it might be before traffic flows again. It may not reduce delay or let one avoid congestion, but it might reduce anxiety.

New kinds of communication and service systems are being introduced to furnish up-to-date information that is more closely tailored to each driver's needs. Similar systems provide information to help find a restaurant or a place to park. There are many schemes and variations of schemes being pursued; some will help and some will disappoint, just as in real life.

While we do not share the optimism put in the pick-another-route-and-go-around-the-congestion dictum, more up-to-date information may help in the management of usage levels to help prevent it from happening. As motorists come to recognize that the information given them about the possibility of congestion is reliable, they are motivated to cooperate with the suggested modifications in routing or believe the advice to put off their planned trip for another X min and, therefore, are less likely to try to second-guess the system.

Accident reduction

Reducing the number of accidents and other incidents that stop traffic could be a great help; some experts say that maybe half of urban traffic congestion is caused by these interruptions. As we noted, metering at on-ramps and smoother and better controlled traffic, in general, help reduce accidents; and since accidents often beget more accidents, the net payoff is significant. Faster response to the accidents that do occur not only decreases the seriousness of injuries but also reduces the time that traffic is impeded and thus the length of the jam that builds up.

Route guidance

Route guidance systems—car navigation systems—have gained in popularity. These are primarily a help to drivers in finding their way, but by cutting down on wasted miles driven, they also appear to have the potential to make at least a slight dent in total miles driven.

In the media these systems are often thrown in the same pot with driver information: "our nifty system will help you find new routes around the congestion." Maybe.

In summary...

Most—maybe all—of the measures noted will help at least a little to improve traffic flow and increase effective road system capacity. In our judgment, however, the gains will not be dramatic nor sufficient to appreciably alleviate congestion in the face of continuing growth in traffic. If we were measuring the improvement between no traffic control and truly advanced traffic control the gains would be very large, but we are not: traffic controls in most cities are already reasonably good, so the room for improvement is less dramatic. The net gains from all these measures may not even be noticeable by the average driver, since the number of vehicles trying to use the streets and freeways are likely to grow at least as fast as the improvements.

Let's now look at some of the other actions put forward to exorcise this old devil.

Better public transportation

We read recently of a poll taken among the drivers of one of the crowded highways in Britain. The overwhelming majority of drivers strongly favored better transit service as an alternative to driving. So far so good. But other questions in the poll brought out an interesting point: few of the drivers had any intention of using it—their motivation for better transit was to get the other drivers off the highway.

But still, the most common suggestion one hears is to provide urban rail transportation and to persuade more people to get out of their cars and use it.

Urban rail transit is "mass" transit—it only approaches making economic sense if there is heavy ridership. There are two primary roles in which this can be obtained. The first is in providing circulation in the core of our larger, denser cities or along dense corridors in such cities. The second is supporting the long commute from outlying, heavily populated suburbs into the core of such cities.

"Dense" is the key word here. As we will reiterate repeatedly, transit is a poor competitor to the car in the low-density cities that grew up around the automobile—Los Angeles and Phoenix, for example. The difficulty is not only that cars are tough competition but that the patterns of travel in these modern cities are too diffuse to generate heavily traveled corridors.

Our older cities—largely shaped around walking, horse power, and early transit—are much more compatible with transit even though they, too, have undergone some of the spatial disaggregation that has followed the faster, more personalized transportation offered by the automobile. Their city centers still offer a high-density core of activities that provide concentrated trip-ends at one end of the transit trip. And driving into this core is usually discouraged both by the high level of congestion there and often inadequate or very expensive parking.

But over time, there are fewer and fewer trips into the center of most large cities; the growth is in the suburbs where both access by car and parking is easier. More and more trips are by people going in ones and twos from many different origins to equally diffuse destinations. Transit has a very hard time serving this kind of unconcentrated flow [8].

New York had the highest transit work trip market share in the United States, 27% in 1990. Washington, DC and Chicago are next, each at 13.7%. In other older transit-oriented cities like Boston, Philadelphia, San Francisco, and Pittsburgh the market share was about 10%. In such cities transit can and does play a vital role in congestion alleviation; if we were to dump its riders back on the highways everything would come to a complete halt.

But holding these levels has been a tough and losing battle. Between 1980 and 1994 inflation-adjusted transit subsidies (to our knowledge there are no public transit systems that are not heavily subsidized) increased by 100%, yet overall ridership decreased by about 10%. Looking at transit-oriented cities, the New York transit market share is 26% less in 1990 than in 1970. Chicago is down 35%, and so on [9, 10].

So we have spent a lot of money attempting to divert more people to transit and the results are not encouraging. Numbers like these suggest that the primary question for big city transit systems is how to gracefully manage a very long decline in markets as well as productivity [11].

Among the many elements of the intelligent transportation system program are efforts to further improve the performance and the ease of use of our public transportation systems. And such efforts are desirable; many people are absolutely dependent on these systems, and there are probably more people who would use them if it were a bit easier and more pleasant. But we doubt that this is the path to significant congestion reduction in our cities.

So we almost have to support transit, but with criteria based more on hardheaded costs versus benefit considerations and less on nostalgic yearning to change other people's travel habits.

The Martian says, *"You are talking about organized transit, the services run by bureaucratic agencies—agencies that claim rights to serve certain markets, lobby for subsidies, follow schedules, and all that. As I look around the world, there are lots more options than that—vans, limousines for hire serving several persons at a time, jitneys, ad hoc car pools, and many other variants."*

(A jitney is a car or van that goes along a route and picks up and drops off people anywhere on that route for a small fixed fare. Others make markets to take folk to grocery stores or visiting. Jitneys have been around for a long time; there are lots of them overseas.)

The Martian is correct. We tend to think public transportation and only imagine something like the big metropolitan transit districts with their streetcar-like, bus, elevated-subway, and commuter services; it blurs our vision of the great variety of non-government-provided services that are so popular in other countries. But rather than plug for the growth of such services, we hear transit folk and politicians urging the development of super agencies, agencies covering lots of territory and all transit operations. Avoiding duplication, efficient service coordination, and other management virtues are claimed. The costs of larger and larger bureaucracies and decreasing innovation and competition go unmentioned.

When we think "group riding" instead of just "transit," the combinations of service are endless. There is sharing the family car, folks going to a ball game who share a car or van, entrepreneurs collecting riders and dropping them off for a fee, for example. The intelligent transportation system products aimed for traditional transit (high-technology fare collection and vehicle schedule keeping and monitoring, for example) miss this market.

Technology can make a huge difference here, too. Already, the cellular phone and computer-aided scheduling are improving group riding services. Such privately supplied services can only be made even more effective with all our fancy new communication and automation capabilities. This market might well be a surprise success story.

Such systems may be very important in providing much better urban mobility for the carless, providing both lower costs and a flexibility of service not available from conventional transit. But while it's wonderful to have another straw to grasp in our assault on congestion, we think this is a weak one for that purpose. First, the size of the potential market in proportion to the problem is too small; and second, it still carries people in relatively small groups over the road system. It's kind of a more flexible carpool, which we will discuss next.

What is the bottom line? The trends say that investments, arm twisting, subsidies, and other things based on the casual slogan "get people out of their cars and into transit"—the services provided by traditional transit districts—will not ease congestion nor serve increasingly diffuse trip patterns.

The more innovative group riding sorts of things aimed at niche markets could very well increase in importance as smart cars, communication, scheduling, and other things come along. Their impact on congestion, however, will be less important than their role in improving urban mobility.

Car pooling

There have been heroic actions to promote carpooling, largely with programs organized by governments or created by firms in response to government mandates or pressures. The thought is, of course, that if folks team up and leave some cars at home, congestion will be alleviated because fewer cars are needed to carry the same number of people, thus reducing delay, energy use, and pollutant emissions. Carpooling is just one of the many ways to achieve more group riding, but the only one being pursued with vigor.

Carpooling has helped some, but, like transit, there is a limit to the blood in that turnip. Carpooling is up against the same fundamental problem as mass transit: living and work patterns are very diffuse and becoming more so, and it is less and less likely that two or more folks have the trip origin and destination and time of travel close enough together to permit pooling. Results have been predictably disappointing.

Think about it: to increase carpooling on the journey to work there has to be at least two or more people who: (1) live reasonably close to a common route to work; (2) work at the same place, or at least places close together; (3) who have nearly the same hours; (4) have jobs that rarely demand variations in those start and stop times; and (5) commute far enough that it's worth the extra hassle and time of picking up the passenger(s) and delivering them back home.

As cities have spread out, carpooling-friendly conditions become harder and harder to meet. The average vehicle on the journey to work has just over one occupant (1.14 to be exact, which corresponds to having 14% of the cars carrying two people or about 10% if half the carpoolers have three occupants).

There is much more carpooling off-peak than during commuting, and overall about one-third of all trips are multioccupancy. On other types of trips people often do things together: shopping, going to entertainment, out to dinner, or visiting. Lots of group riding is by people belonging to the same household. The average occupancy offpeak is 1.6 people per vehicle-mile, compared to 1.14 during peak hours. The downside is that this higher occupancy is attained when congestion is not at its worst, when we really need it.

Adding to the trend of the spreading of settlement patterns are changes such as smaller living groups, which probably reduce vehicle occupancy for recreational, visiting, shopping, and other trip purposes of households; more persons in the population working; and increases in specialized activities at trip ends, which may decrease the motives for group riding.

Just looking at trends we might conclude that group riding is on the way out, and that could well be true overall. But the more flexible entrepreneurial systems mentioned earlier, helped by the communication, scheduling, and other new technologies coming along, should increase the quality and competitiveness of both old and new services—airport vans, ad hoc journey to work chauffeuring, and car pools serving after school sports.

Don't expect too much from carpooling, but don't write it off entirely.

Reversible lanes

Redesignating the permissible direction of traffic on lanes to accommodate the dominate flow is an old idea. And it's a very good idea, but sometimes the geometry of the road or freeway does not permit it, and the schemes for making a quick reversal and for maintaining adequate safety in the meantime have been limited.

There has been considerable progress in overcoming these difficulties, often starting in the original road design. A technique called "zippering" was introduced many years ago on the Coronado Bridge in San Diego. In this scheme a device looking like a giant caterpillar straddled the concrete traffic separators in the middle of the bridge and, as it moved forward, shifted them from one side of the lane to another, thus taking away a westbound lane and making it eastbound or vice versa.

The application of this scheme to much longer stretches of freeway was awaiting faster "zippers"; the original ones were too slow to "rezip" long stretches in a reasonable time.

There are other schemes: posts on little elevators built into the roads so that all those on one side of a lane can be lowered into the road and those on the other side raised, thus moving the barrier from one side to the other.

High-occupancy vehicle lanes

What about high-occupancy vehicle (HOV) lanes? These are one of the incentives offered to encourage more group riding by providing special, uncongested lanes to those who carpool.

Dr. Joy Dahlgren of the University of California at Berkeley's Institute of Transportation Studies has modeled a wide variety of HOV-usage situations. She found that unless carpooling is already very heavy, adding a HOV lane worsens congestion rather than helping it. In her assessment, doing away with HOV lanes altogether is almost always the preferred strategy [12].

How can you tell if an HOV lane is paying off? Just look at the usage of that lane. If it's very heavy, if the lane is full or nearly full, then it is paying off. But if it's usually half empty, then there would be less total delay if it were converted to an ordinary lane.

Vehicle automation

In the much longer term, the automated vehicles discussed in the smart car chapter offer the potential to double or triple freeway capacity. Automation will also in time significantly increase the potential throughput of intersections on surface streets.

As we already noted, automated systems have much faster and more reliable reflexes than human drivers, and therefore automation lets vehicles drive closer together safely; that is, they allow higher vehicle densities without congestion. "Platooning"—the electronic entraining of two or more vehicles that we discussed earlier—raises densities still higher because it decreases still further the average distance between vehicles. The flow capacity in a lane can increase by a factor of two, and it is not unreasonable to think that we may achieve a factor of three. Our four-lane freeways in the future will be able to carry the equivalent of eight to twelve lanes today.

These huge increases in effective capacity sound wonderful, but remember that will take at least several decades to fully come to pass. They depend on having a large proportion—the more the better—of the vehicle fleet equipped with intelligent cruise control, including a platooning capability.

But we won't have to wait for 20 or more years to start enjoying some of the benefits of automation, such as intelligent cruise control; sensors that tell us what is in blind spots; better brakes; and a whole list of other things that ought to lead to smoother, tighter, safer driving—all long before we get to full-scale platooning.

So while we can anticipate that automation will offer real benefits a few years down the road, we still must deal with our problems now.

Other favorites

Work-hour staggering among the various organizations in an area spread the peak-hour commute and, thus, lowers the traffic level. This can be done in a place like Washington, DC, where a high percentage of employees work for one organization (called government), but is harder to pull off in other places. Government is the 800-lb gorilla, and government agencies can afford to be out of step with their customers and suppliers. Most businesses can't.

There are other common-sense actions that can help, like making sure there is plenty of parking downtown so that cars don't have to keep circling the block looking for places. Pedestrian crossovers above the streets are a big help. Minneapolis, Tokyo, and Seoul are examples of this.

Scheduling routine road maintenance activities to keep them out of the rush hour is an obvious step. The principal ingredient here is common sense and the will to apply it (and maybe the money for premium pay).

Rather than go through today's long list of demand management schemes, we will just say that in our judgment the impact of most of them is too small to matter much. But there is still a "clutch at any straw" attitude on the part of the people who are trying to cure congestion today, so many are pursued, often vigorously.

Where does all this leave us?

The answer to that devil congestion is not here.

We seem to have decided that we will live with congestion as a normal part of life. We rail about it; and spend lots of money on traffic management, transit, driver information systems, and other techniques that have some marginal payoff; and even build some roads, but we really don't expect to scourge this devil.

The only technique that seems to hold much hope in most people's eyes is congestion pricing. We have the technology to automatically collect a toll anytime anyone drives onto the freeway or enters a busy section of town. The thought is that if this price is set high enough, then usage will drop to be within the capacity of that freeway or subregion. It has been done in many parts of the world, and it works insofar as congestion alone is concerned. We will talk about this more later; for now its enough to say that at least one reason we don't use it is because most people seem to prefer the devil to the pain of this particular cure.

What is the downside of this kind of muddle-along strategy? Simply that by letting congestion persist we reduce the potential capacity of the road system and suffer all the other penalties of lost time, more fuel consumption, and a reduced quality of life in the city.

There are other options, which we will discuss in the next chapter.

References

[1] *Highway Capacity Manual*, Special Report 209, Washington, DC: Transportation Research Board, National Research Council, 1994.

[2] http://www.its.dot.gov, Website of Joint Program Office of the Intelligent Transportation System program, U.S. Dept. of Transportation, Washington, DC.

[3] http://www.itsa.org, Website of ITS America Organization, Washington, DC.

[4] http://www.itsinternational.com, Website of *ITS International, J. Advanced Transportation Infrastructure*, Route One Publishing Limited, London.

[5] "Ramp Up the Volume," *ITS International, The J. Advanced Transportation Infrastructure*, Nov./Dec. 1997.

[6] TOLLtrans, supplement to Traffic Technology International, Oct./Nov. 1998.

[7] "Tolling," *ITS International,* op. cit. Jan./Feb. 1999.

[8] Pushkarev, B. S., and J. M. Zupan, *Public Transportation and Land Use Policy,* Bloomington & London: Indiana University Press, 1977.

[9] Taylor, B. D., and W. S. McCullough, "Lost Riders," *Access*, Vol. 13, Fall 1998, pp. 27–31.

[10] Pucher, J., T. Evans, and J. Wenger, "Socioeconomics of Urban Travel," *Transportation Quarterly*, Vol. 52, 1998, pp. 15–33.

[11] Lave, C., "Measuring the Decline in Transit Productivity in the U.S.," *Transportation Planning and Technology*, Vol. 15, 1991, pp. 115–124.

[12] Dahlgren, J., "High Occupancy Vehicle Lanes: Not Always More Effective Than General Purpose Lanes," *Transportation Research*, Vol. 32A, 1998, pp. 99–114.

5

Congestion: A Better Strategy?

If one is trying to solve a problem, it often helps to understand it.

Here we lay out what we consider to be a better strategy for congestion prevention—not 100% prevention, but certainly a significant reduction in the frequency of its occurrence.

We start with a short description of the basic nature of freeway traffic flow, why traffic behaves the way it does, and what makes it decide that it doesn't want to flow smoothly anymore and starts to congest.

It's quite possible that we'll tell some of you a bit more than you really wanted to know, but we drag you through it for two reasons. First, it is the nature of the phenomenon that leads us to the specific prescription we offer. Second, the next time you are creeping along on your favorite congested freeway, it will give you something to do to watch the traffic flow dynamics and see if you think we've got it right.

Freeway Traffic Flow 101: The nature of the phenomenon

Consider a freeway lane. The maximum number of cars the lane can hold is around 300 VPM; this is the number you get if they are just parked there, with, say, only 1 ft between them. But no one can drive with cars this densely packed, so the flow would be zero.

Let's take half the cars off, reducing the density to about 150 cars per mile. This leaves about an 18-ft gap between each of the remaining cars. People are now able to drive, but they are not willing to go very fast with the car in front this close. In fact, with this vehicle separation, people are usually not willing to drive much above 12 or so mph on average. At 11 mph and 150 VPM, the flow is 1,650 vehicles per hour per lane (11 multiplied by 150), which is about three-fourths or so of the maximum flow capacity of 2,200 vehicles per hour per lane we have assumed for our particular freeway.

So let's again take half the remaining cars off our freeway lane, cutting the density to 75 VPM. This opens the average gap between cars to roughly 55 ft. It turns out now that drivers are willing to speed up to 26–30 or so mph. At 28 mph the flow is 2,175 vehicles per hour per lane, nearly the maximum of 2,200 (for our illustrative example).

Now if we take a poll of these drivers, they will complain that the freeway is congested because they can't go fast enough. A traffic engineer might admit the service quality (the traffic engineers' term for ease of travel) is down a bit but point out that the freeway lane is carrying pretty close to as many vehicles per hour as it is capable. But discontented drivers are not the compelling reason for not wanting to operate at this vehicle density. We'll show why later.

We hit the maximum capacity of 2,200 vehicles per hour per lane on our illustrative freeway at just about a density of 50 vehicles per lane and a speed of 44 mph. The drivers are still not totally happy because they would prefer to drive faster. We can't say that we blame them.

Now let's reduce the vehicle density still further to 30 VPM per lane. The traffic speeds up to 70 or so mph as there is now nearly 160 ft between vehicles. All the drivers rejoice, even though the flow in vehicles per hour actually drops slightly to about 2,100 vehicles per hour, because the increase in speed is more than counteracted by the decrease in density—the numbers of vehicles per mile moving at that speed. Lower the density still further and the flow in vehicles per hour decreases even though the drivers can now go even faster (although such is frowned upon in some circles).

You may have already deduced that the relationship between speed and vehicle spacing (or density) is determined purely by a driver's behavior: how fast the average driver is willing to go with a given space between his or her car and the one in front. And since density and speed determine flow, the whole phenomenon is based on how people behave. We try to illuminate this behavior and its impacts a bit more in Appendix A.

Behavior changes. People drive differently when there are curves or merges, or when it's raining, or at night, or where there is a police cruiser nearby, or if there is an accident somewhere in sight, for example. (Sometimes not differently enough, hence the rise in accident rates in bad weather.)

Nor are freeway lanes all alike. And the habits of the driving populace may also differ from city-to-city. So remember, the numbers we have used in our little illustration are purely for that.

What's magic about a vehicle density of 50? Why is a density of 30 VPM per lane preferred to 75 VPM per lane? The drivers, of course, prefer the 30 because they can drive faster, but the flow capacity is actually a bit lower at 30. The main reason is, in the jargon of the professional, when vehicle densities are above our magic value of 50 the flow is "unstable." This is a crucially important aspect of the phenomenon, so we momentarily dwell upon it.

The reader may have noticed that lane flow peaked at a vehicle density of 50, rising to that value as density increased, even though drivers were having to slow down because the spacing between vehicles was getting smaller. We call densities below 50 "stable." Below 50, as density increases (spacing gets smaller), drivers slow, but flow in vehicles per hour keeps rising: increases in density more than compensate for losses in speed. These relationships are illustrated in Figure 5.1.

But when vehicle density rises above 50 VPM per lane the opposite happens: if density increases, drivers slow, but flow decreases: on this "unstable" side the speed drops faster than density increases. We will illustrate what this implies for traffic behavior with a couple of examples.

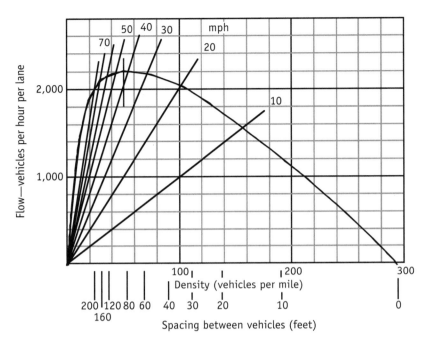

Figure 5.1 A highway capacity curve for uninterrupted flow.

Traffic is rolling along at 70 mph with a vehicle density of 30 VPM per lane, so the flow quantity is 2,100 cars per hour. At 30 VPM per lane we are below our magic 50, so we are in a stable operating region.

Now more cars enter the freeway from an on-ramp. With more cars, the spacing between cars is reduced a bit; and since drivers on average are a little uncomfortable going 70 with this closer spacing, they slow slightly. But, Eureka! Freeway flow capacity does not go down, it actually increases, albeit only slightly, because the higher density of vehicles more than compensates for their lower speed.

Drivers may be slightly less happy, but from the point of view of the freeway there is no problem. Now more cars can flow away from the vicinity of the on-ramp every minute than are flowing into that segment of freeway. Traffic does not jam.

Now let's picture the situation with the density at 75 cars per mile per lane instead of 30; here we are in the unstable operating regime. Traffic is slow, only 28 mph. The following is what could happen. Assume more cars enter the freeway from the on-ramp. Just like before, the spacing between cars now has to decrease a bit. A few drivers brake slightly and traffic therefore slows. But now the traffic flow rate goes down, not up, because the speed drops by more than the density increases.

At the lower flow, fewer cars are flowing away from the vicinity of the on-ramp, but the same number are entering that segment of freeway from upstream. With fewer vehicles leaving than are entering, cars are forced even closer together. So drivers slow still further and flow drops even more. Traffic slows even more. Just following the mathematics, this process doesn't stop until traffic stops completely.

If traffic really behaved this way all the time, if it were really this unstable, we would have beaten congestion long ago, because we couldn't have afforded not to. In a world where every time the density exceeded the peak flow density all traffic stopped, we would have had the incentive to strictly control vehicle density and nobody would complain about having to wait in line to get on freeways.

In real life, freeway traffic doesn't behave like a mathematical formula. We don't know the mental calculations drivers go through to determine how closely they are willing to follow the vehicle in front: it seems to be an instinctive judgment based on "I want to be able to stop if he does." Drivers are often willing to drive a little closer for short periods, so they don't always slow down when distances between vehicles decreases slightly; they seem to be willing to trade increased vigilance for smaller following distances. In any event, this variable behavior introduces a kind of partial "forgiveness" in traffic flow behavior on the unstable flow side of traffic operations.

But there is not total forgiveness. Sometimes this slow down, coast, brake does lead to traffic coming to a complete halt. But often it doesn't. Often there are stretches of lower density traffic into which our incipient jam can expand before traffic becomes seriously congested.

So densities beyond our magic number of roughly 50 can often be tolerated, but increasingly it becomes a gamble. As vehicle density increases further, the chances also increase that any upset to smooth traffic flow can turn into a jam: cars entering from an on-ramp, changing lanes, or some drivers deciding to slow down to gawk at something. Perhaps the metaphor is too strong, but heavy traffic shares some characteristics with Russian Roulette.

It will also help a lot to eliminate accidents. Now eliminate is a strong word, but there are certainly things that can be done to reduce them. We will talk later about how we might accomplish these things.

A freeway game

When you are driving along the freeway you obviously can't count the cars and therefore determine vehicle density around you. But you can make a good guess. If you are speeding along at 60 or 70 mph, you can be pretty sure that the densities are well below the peak density of your particular freeway.

But if traffic has slowed to somewhere around 30 or 40 mph and you feel uncomfortably close to the car ahead, then traffic is probably pushing the limit. Watch the tail lights to see if there are lots of flashes of red from brakes being tapped. Then watch for disturbances to flow, like a few hot-shots changing lanes or cars entering from on-ramps. You can expect more flashing brake lights and a bit more slowing of the traffic stream. If you're lucky, flow will recover; if you're not, then you might well get into that mode of complete stop, creep up to 15 mph, and then stop again. This will give you plenty of time to really analyze the phenomenon.

So once again—the prescription

We laid out the common-sense prescription in the last chapter:

> First, to increase road and freeway capacity as much as we can within the many constraints of real life; and then control, regulate, or otherwise manage to keep usage levels within that capacity.

The first is what we have been doing with all our fancy traffic controls and many of the other functions we described in the last chapter.

The second part of the prescription still needs filling. Our analysis of the flow phenomenon gives us a much more precise idea of both the why and the what.

One important implication is that it is not enough to keep the average usage (i.e., vehicle density) over the course of a day below the critical value for that freeway or to average the density over the whole freeway; we want to keep the vehicle density below the critical density at every instant over every section of the freeway. Letting it momentarily get out of hand a little at one point doesn't automatically spell congestion, but the probability that it could spell congestion goes up and the worse the deviation and the longer the time it persists and the larger the section of freeway over which it is happening, the greater the probability that congestion will, in fact, set in.

And once serious congestion starts, the game is often lost: traffic piles in from behind, the jam builds up, and large jams can sometimes take hours to clear.

What this suggests is that techniques to just lower the average usage level aren't sufficient unless they are terribly draconian because they can't prevent chance fluctuations from exceeding our desired control limits.

What, then, to do?

The first thing we would do is put a big sign on the wall of every traffic control facility:

WATCH VEHICLE DENSITY!

The other thing is to figure out how to do something about it. We think that stringent, system-wide ramp metering is the primary tool for accomplishing this on freeways. As we described in the last chapter, ramp meters have been around for a long time, but in most places they are not being used as effectively as they could to control vehicle density and therefore the probability of congestion.

There are two basic reasons this strict control over freeway density with ramp metering hasn't happened more widely already.

First, it is not easy to do. Effective access control over the whole freeway system in a reasonably large city (where most of our problems are) is not a simple job.

Given that we have the instrumentation to know what the vehicle density is in every segment, what do we do to control it? As we noted, a vehicle entering the freeway now will in a few minutes be miles downstream. So what happens at point A now will partially determine the traffic conditions at point B later. The system has to think ahead, so to speak.

There are complications. That vehicle that got on five minutes ago may have left the freeway two minutes later, or, if upstream of a point where the freeway branches, may be on any of the branches. But this is not a fundamental problem: a knowledge of the average patterns of traffic can be used to reduce much of this kind of uncertainty.

Nor can metering solve every problem. Once traffic is on the freeway, it is too late to exclude it with metering. This problem can be particularly sticky if heavy flows are entering the metered portion of the system from outside so that the freeway is essentially already full before metering can become effective. There are ways to alleviate this, but it does complicate the picture.

As we noted, metering systems do reduce accidents by spacing the vehicles entering the already flowing traffic, but no metering system can totally prevent accidents nor the congestion that results. But they can reduce the size of the jam using the ramp meters to immediately prevent more vehicles from entering the system upstream of the accident. Control of the speed of the upstream traffic already on the freeway could also help by slowing the rate at which vehicles reach the jam.

All this is to make the point that setting up a coordinated freeway density control system is not a casual undertaking. But we think it can be done. Many cities already have the ramp meters in place; they are just not usually being operated to accomplish the goal of ubiquitous and continuous vehicle density control.

The second basic problem is that there is the very real likelihood of local resistance to actually doing it. We've all seen that at times ramp meters generate long waiting lines at the on-ramp; this typically occurs at peak hours when the freeway is badly congested. Motorists complain because they do not see the reason they have to wait in such lines, and the local jurisdictions complain because the lines back up and plug the local streets.

The perception is that more stringent metering will cause waiting lines to be even worse.

We think this is at least a partial misreading of the phenomenon. The underlying fact is that the congestion on the freeway is helping make the lines worse. If we eliminate this congestion, then cars could enter the freeway faster than before.

For example, if the freeway is flowing 50% better than when it was congested, then cars can enter 50% faster, and, other things equal, lines will be shorter, not longer. But the waiting line problem and this "dumping congestion on us" complaint is not something that we can brush off lightly. And we don't, we address it below under "The Waiting Line Bugaboo."

We will never be perfect; there will always be something—maybe a broken water main or the traffic accident we talked about—that we can't foresee and totally prevent, but we can prevent having just volume of traffic itself sow the seeds of its own impediment.

Off the freeway—surface streets

We think the same principles of strict control of local vehicle density apply to arterials and surface streets as well as on freeways.

The dynamics of traffic behavior on ordinary city streets and arterials is much less influenced by the flow phenomenon just described than it is by the traffic at intersections. But common sense tells us that the same general prescription used to control overall vehicle density should work here, too. Just think about it: when there aren't many vehicles in a given section of the network, traffic usually flows reasonably well. But keep adding vehicles and at some point the intersections stop clearing at each cycle of the traffic lights, queues start growing, and finally everything grinds to a halt.

The general idea is to subdivide the urban area into subareas and determine the number of vehicles that each subarea is capable of accommodating without causing congestion. The prescription here is to monitor the number of vehicles in each area and to use a combination of traffic control and demand management to keep the density below the critical value.

How can traffic control help? One way might be to use the traffic lights as metering devices—oversimplifying a bit, to bias the traffic lights to longer reds for traffic approaching areas where density is reaching critical levels and to bias them toward more green for those going away from these higher density areas. Here is where information to drivers should also help. It would take some trial and error to work out real-life techniques; they are not a normal part of traffic control systems. Some special treatment of arterials might be required.

This might be a good time to also see what can be done to make each subarea capable of handling even more traffic: is there enough parking, so that part of the traffic is not just cars running around looking for spaces, and are there things that can be done to decrease the impact of pedestrians on flow, for example.

The waiting line bugaboo

Eliminating congestion with an advanced metering system will increase effective freeway capacity and let vehicles enter the freeway faster, therefore potentially shortening, rather than lengthening, any waiting lines at on-ramps.

But not forever. There is always latent demand waiting in the wings; these are people who have postponed trips or changed their travel habits to avoid the congestion. When the congestion is removed, they will start to indulge their desires. Over time both this unleashed latent demand and general growth and expansion of the area will increase desired usage beyond the available capacity of even uncongested freeways. As demand increases at peak hours the queues at on-ramps will become longer and more chronic. Even before then, there will still be long queues at times, as when there is an accident and congestion from the resultant blockage slows flow.

Then the bugaboo of ramp metering will become long waiting lines at on-ramps.

In a sense, we are using congestion today to allocate scarce road space: "if you must travel, then get out there and suffer; if not, then postpone or forego the trip." If we turn off the pain-of-congestion, then the queues themselves pick up the role of being the motivation. The pain-of-congestion is replaced by the pain-of-waiting-in-line.

Waiting is onerous, even if the driver is finally able to get on a free-flowing freeway and even if the traffic engineers assure everyone that the freeway is carrying more cars than it ever did back in the old days when it was congested. An information system that tells drivers the waiting time they should anticipate at different freeway access points should be helpful to people who have some flexibility in scheduling their trips. Such a system would permit drivers to more intelligently modify their behavior to reduce their own inconvenience and costs.

Road pricing, the economists' favorite

At some point, nearly every urban area will have a choice to make: either build more roads and freeways, or put up with the longer wait times to get on freeways and into the denser portions of the city, or introduce another way to allocate the too short supply of effective road system capacity.

We talked in the previous chapter about many of the demand management techniques now in use and concluded that there's not much more help available there. So we come back to the idea of using money—price—as the basis for allocation, just like we allocate who gets Cadillacs and who gets Chevrolets, who gets steak and who gets Hamburger Helper.

The idea is to charge each vehicle for its use of the freeway (and perhaps other areas of high chronic congestion) during periods of peak usage. If the peak price is set high enough, then people will shift their habits enough to reduce the gap between usage and capacity.

A fundamental problem with road pricing is that people view their cars as their freedom and don't like any actions that they perceive limit this freedom to go when and where they want to go. They also say that they have already paid for roads through fuel and other taxes. Many people see road pricing as just another clever scheme for the government to collect more money: "why should we have to pay again for something we have already paid for?" Many say they are concerned that it seems unfair to low-income travelers in spite of possible subvention by subsidies.

These sentiments, right or wrong, make strong headwinds for road pricing. So almost any action to limit usage takes some explaining, and the benefits have to be persuasively clear in order to gain acceptance. The disease competes with the cure.

How high need the price be set? Many argue that it should reflect the costs that the congestion causes, which is why such schemes are often termed "congestion cost" pricing schemes. The car entering traffic should be charged the congestion cost it imposes on other cars.

Some go beyond congestion costs and ask that cars pay for pollution and other costs imposed on others—external, nonmarket, costs. Or add a little more to cover the costs of improvements to facilities. One might treat it as a business and charge what the traffic will bear—extract as much as people will pay. That is a scheme with the fancy title "reservation prices"—imagine making drivers bid for space on the road!

Most of these prices would get pretty difficult to calculate without considerable controversy.

A pragmatic approach is to set the price just high enough to accomplish our goal of roughly matching demand for travel with supply (road capacity). We will find the right prices by trial and error, with hopefully not too much of either.

There is the problem of how to do it. The idea of a toll booth on every freeway on-ramp or on the main arterials leading into downtowns seems a bit ridiculous. But technology to the rescue: one of the most important innovations of the last few years is a way to collect tolls automatically without the vehicle having to slow down. We've already mentioned these systems, which are already in use in a number of sites throughout the country and the world.

In our judgment, the logic behind the idea of road pricing is sound. Road space is a resource in short supply. Road pricing rations these resources according to how folks value them: what they are willing to pay. We can use a subsidy to help out the less affluent, as we do for many other things. Road pricing can make a positive contribution to making travel more pleasant and predictable by cutting waiting lines and can help keep surface street usage below congestion points. Money collected from tolls can be used for good things, including building more capacity where the environment permits.

We think it appropriate to raise one red flag. Congestion pricing is the darling of transportation professionals and economists, and it could very well absorb resources of time and attention for several decades. Viewed as the magic bullet, pricing could so absorb attention that other companion actions that could make it more effective, like advanced metering, could easily fall between the cracks.

We discuss next how, magic or not, congestion pricing could be a more effective bullet if it were to be combined with the advanced metering systems we discussed earlier. In fact, the sensible way to collect the congestion pricing tolls on freeways is to combine that function with the metering system.

The synergy of metering plus pricing

The purpose of these demand management schemes is to lower the general level of vehicle usage of the freeway and street system, particularly at peak hours. It is our thesis that such measures will be most effective used in combination with the advanced metering systems and area density control schemes previously discussed; there is important synergy between the two.

The drawback to total dependence on demand management for congestion prevention is that such measures tend to be blunt instruments: they can pull down the average levels of usage, but they cannot prevent chance fluctuations in traffic levels. And the congestion phenomenon is such that even short-lived increases in vehicle densities in a freeway or roadway above the effective capacity increases the probability of a jam occurring. And once a jam occurs if traffic is heavy, it grows quickly and dissipates slowly.

The role of advanced metering, then, is to prevent such fluctuations from causing congestion (and to prevent or slow further inflows into accident sites). By doing so, the demand management measures can be more relaxed because a higher level of average usage is permissible. This is simply because pricing's role is no longer primarily to prevent congestion—which would require that average usage be kept low enough that fluctuations into the congestion region are rare—but becomes that of keeping queues reasonable at on-ramps and in areas of high traffic density. This lowers the prices that must be charged to keep traffic flowing.

Looking ahead: a summary

How to cope with continuing traffic growth?

We suggest that reduced congestion could be a big step in increasing effective capacity of the existing system. We already noted that congestion is now robbing us of perhaps a third to a half of our available freeway capacity at peak hours. Cure it, and we get it back. Advanced metering is the first line of attack.

Because significantly alleviating congestion increases available capacity, it may be that in the near term more demand management measures will not be needed. But in time latent demand and new demand stemming from economic growth will exceed the gains from congestion relief.

From that point—when the maximum effective capacity of the road system is reached, when transit can't meet increasing need, and when waiting lines reach the limit of driver tolerance—further mobility cannot be supplied without building more roads or increasing the number of lanes available in those we have.

If we cannot expand the road system adequately, then the issue becomes how to allocate that which is available. We already discussed the options and emphasized the basic logic of moving to road pricing. We also discussed how, when this stage is reached, the combination of advanced metering and road pricing is superior to either alone.

Road pricing may be a hard sell—it seems to fly in the face of our almost sacred right to use the roads whenever we want without waiting for anything. But reality is reality, and road pricing may make a lot more sense than having a city choke on its own congestion.

Further increases in personal mobility will, as far as we can see now, have to await the automated vehicle—and perhaps a new kind of city. We will explore this latter possibility in the next several chapters.

Selected bibliography

Gomez-Ibanez, J. A., "Pricing," *Essays in Transportation Economics and Policy*, J. Gomez-Ibanez, W. B. Tye, and C. Winston (eds.), Washington, DC: Brookings Institution Press, 1999, pp. 99–136.

Johansson, B., and L.-G. Mattson (eds.), *Road Pricing: Theory, Empirical Assessment and Policy*, Boston, MA: Kluwer Academic Publishers, 1995.

Vickery, W., "Pricing in Urban and Suburban Transportation," *American Economic Review*, Vol. 53 (Papers and Proceedings), May 1963, pp. 452–465.

Walters, A. A., *The Economics of Road User Charges*, Baltimore, MD: The Johns Hopkins University Press, World Bank Occasional Papers #5, 1968.

Ward, J. D., "An Analysis of the Requisites and Strategies for Urban Congestion Alleviation," in support of ITS Architecture Development Program, U.S. Dept. of Transportation, Washington, DC, 1995 (unpublished).

Appendix 5A:
Variable driver behavior and the highway capacity curve

The purpose of this appendix is to illustrate how driver behavior affects the highway capacity curve: the relationship among speed, density (vehicle spacing), and flow in a traffic stream.

The effect on which we focus here is the driver's decision of how closely he or she is willing to follow the vehicle in front without slowing. To generate the curves we used the following model.

We assume that Car B follows Car A such that if Car A suddenly brakes to a full stop (or is caused to slow by some other mechanism) Car B is sufficiently far behind that it is able to also brake to a stop without hitting Car A: safety is maintained. We assume that Car B only brakes after some time delay, which we might think of as its driver's reaction time.

For our illustration we assume that Car A is going to brake at 0.7g's (22.4 ft/s^2) and that Car B, after a braking delay, brakes at 80 percent of that level (0.56g's, or 17.92 ft/s^2).

Both values represent vigorous braking, well above that of ordinary driving. While this choice of values is somewhat arbitrary, when combined with braking delays that roughly represent human capabilities, they do produce highway capacity curves that seem to reasonably approximate some observed traffic behavior. We doubt very much, however, that drivers explicitly use such mental models as the basis for their instinctive judgments of proper following distances at different speeds.

Braking delay is our proxy for how closely a given driver is willing to follow the vehicle in front. A short braking delay characterizes the driver who is willing to follow closely, at least implicitly assuming himself or herself to possess reflexes capable of fast reaction if the need arises (i.e., the car ahead slows rapidly). Longer delays characterize drivers who are less comfortable following closely and, therefore, chose to maintain a larger gap between vehicles.

Figure 5A.1 lets us illustrate the effect. Vehicle flow in vehicles per hour is the product of speed in miles per hour times the density of traffic in vehicles per mile, so the flow is a straight line as shown for any given speed of traffic. Now if we could extend these lines out to the point where

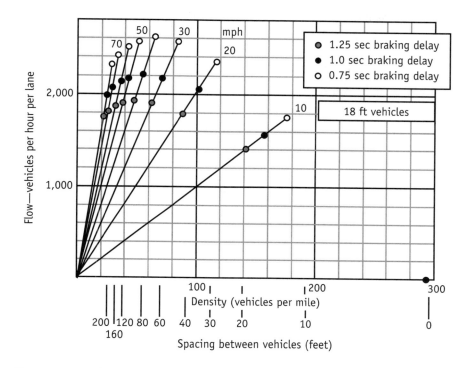

Figure 5A.1 Flow versus vehicle density for different speeds.

cars were bumper to bumper, which, for the 18-ft vehicles we assumed here, would happen when there were 293 vehicles crammed into every mile. At 50 mph, for example, we would have a total flow of 14,665 vehicles per hour per lane!

But people spoil the whole thing. They seem to get nervous about driving at 50 mph bumper to bumper. In fact, they get nervous a lot sooner than that. As the density of traffic increases, and therefore the gap between vehicles decreases, the most cautious driver (that we characterize with the 1.25-sec braking delay) would refuse to continue at 50 mph when the gap narrowed much below 122 ft (a density of 38 vehicles per mile); this point is the little shaded circle on the 50-mph line in Figure 5A.1. At this point our driver would begin to slow. By the time density reached 48 VPM, the gap would have narrowed to a little over 92 ft and our cautious driver would have slowed to 40 mph. If all the drivers were exactly like our current hero, the total flow would have actually increased from 1,890 to 1,910 vehicles per hour.

Now those drivers who think they have faster reflexes are willing to maintain their 50-mph speed a little longer, as shown by the other little circles on the 50-mph line. But they, too, finally throw in the towel and begin to slow as density continues to increase.

This is the phenomenon that defines the highway capacity curves, as shown in Figure 5A.2. But note that in real life the curve or boundary depicting the maximum flow at a given speed is not a precise and invariant line because it varies with the aggregate behavior of the drivers at a particular time. Not only are all drivers not the same, but any individual driver doesn't follow the same pattern all the time. And all kinds of factors can affect this behavior. Maybe the highway capacity curve should be renamed the highway capacity fuzzy regions.

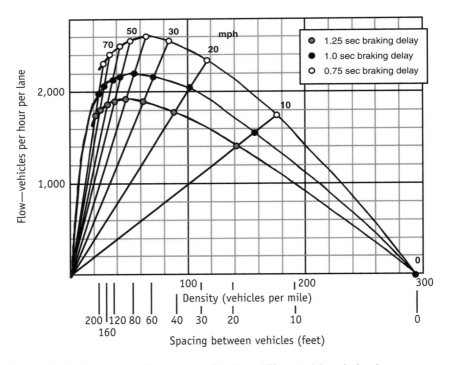

Figure 5A.2 Lane capacity curves reflecting different driver behaviors.

6

Some New Kinds of Cars

The age of specialization is upon us.

The first part of our story is about empty seats, specifically the three empty seats that are in at least 86% of vehicles being used in the daily commute. Do we need all these empty seats in every one of the family's cars? Do we need all that space for vehicles? Those questions have been looked at before, but that doesn't preclude our looking at them again [1].

The commuter car

Imagine this: a single occupant vehicle with room for another person behind the driver in a pinch. With a single occupant, the car can be half-width, that is, one half the width of a conventional car. With streamlining and a smaller cross section being pushed through the air, aerodynamic drag would be significantly reduced. With an empty weight of, say,

700 lbs and suitable aerodynamics, a 20- to 30-horsepower engine would give good performance at 80 to 100 miles per gallon.

Let's shoot for, say, a $5,000 vehicle. But the market might ask for something fancier, so 1,000 pounds and $10,000 might be in order. Who knows? Some innovations take off at the expensive end of the market, beginning as expensive novelties. That might be the case here.

That 700-lb specification is not so silly. It can be achieved with conventional materials. Except for the half-width, the specification is about what Ferdinand Porsche had in mind when he set out in the early 1930s to design what became the VW Beetle. Porsche saw that the Beetle could be built just like an ordinary car, only smaller. He needed a more fundamental approach. His design was different enough from conventional designs that he was able to reach the goal of about a $400 vehicle in then dollars. His VW was fuel efficient and suited to the proposed autobahns.

To capture the high-performance idea yet avoid a vision of small conventional automobiles, we will say "commuter car." That description suggests one function of the vehicle we imagine, commuting; and it also says that range, velocity, easy parking, cost, small turning radius, and some other things are important vehicle attributes. It would be cheaper, faster, better in that market niche.

A narrow, half-width vehicle raises the question of stability. The center of gravity could be made very low, but this would create difficulties in getting in and out of the vehicle and in seeing what is ahead and around and in being seen by other drivers. As an alternative to low seating, the trick would be to design a vehicle that leans like a motorcycle, thus permitting normal, upright seating. General Motors has a prototype able-to-lean vehicle called the Lean Machine (two wheels in the back, one in front), and we have seen proposals for a number of similar vehicles [2]. The leaning gives stability and a fun-to-drive character to the vehicle. Remember cornering on a bicycle, leaning into the curve? The leaning vehicle would give that feeling. That was more fun than today's sliding across the seat and hanging onto the steering wheel that occurs when a standard car corners at speed. Figure 6.1 illustrates the Lean Machine and motorcycle-like leaning.

Sometimes one hears about microcars, and we think of those available on European and other markets as small big cars. They are made up of

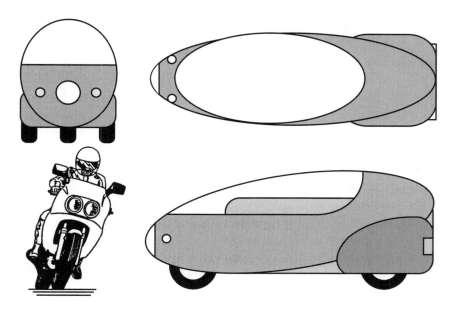

Figure 6.1 Comparing the Lean Machine and a motorcycle [3].

10,000 or so parts and are assembled and marketed as large cars are. To keep costs down, they may not have the "flash, polish, and bells" of large cars. From a mechanical point of view they are just small, cheap versions of large cars.

Porsche aimed for the family car, two adults and two children, for the single car family. Nowadays, there are lots of families that have a car for every driver, and many have sports, vans, small trucks and other specialized vehicles. A commuter car or some other car is imagined as another in a collection of specialized vehicles. It would be a car especially suited for one-person travel where good performance and maneuverability are desired.

Cheaper, faster, better? We said cheaper to own and operate; and very parsimonious in its use of gasoline. Faster? We imagine a high-performance vehicle that could scoot through traffic as a motorcycle does—moving between lanes in slow moving traffic. Better? That is for the market to say; we assume for now that it says it is. We'll talk about that later.

On a different track: the neighborhood car

Imagine a park bench, sofa, or love seat with wheels (well, maybe not a love seat). If the sun is beating down or if it is raining, a cover would be nice. Cold? Let's enclose it. Want to use it for transportation? We need a power train—batteries and electric motor or internal combustion engine—and a platform for your feet.

You already know about vehicles like this. Perhaps after registering at a resort in Hawaii or Arizona, a golf cart-like vehicle toted you and your companions from check-in to your room or suite. That service is not a surprise for tourists. It is found in many resort areas; and such small cars are used on golf courses, in airports, and at other places. In Arizona, Florida, and other states offering warm weather, retirement, and golf course situations, such cars are used not just for golfing but also for a variety of short trips.

Using golf car-like vehicles as cars to get around the neighborhood is cheaper than using a conventional automobile, faster than walking, and better in fitting into the local environment. It is energy efficient, too. Operating at slow speeds avoids much of the energy cost of aerodynamic drag—the drag you feel when you hold your hand out of the car window at speed. It is the drag that the engine in your car works so hard to overcome as you cruise along.

Looking ahead to relatively more older folk living in retirement environments as well as to the increasing popularity of golf, one might suppose that there might be a modestly growing market. But it isn't big, being constrained mostly to the retirement-community-in-great-climate market.

This is not a new idea. In fact, there are already neat little battery-powered, golf-car-derivative enclosed vehicles on the market, vehicles that are built to conform to the minimal safety standards specified by the Department of Transportation's National Highway Transportation Safety Administration (NHTSA). Some states and NHTSA have asked for registration plates and lighting.

But such a simple vehicle wouldn't be so useful where it is very cold or very hot for much of the year. Conventional, relatively inexpensive batteries do not perform well in cold climates and would not be a good energy source for car heating. That, or the need for air conditioning, might imply a new design, or may be motivation for powering the vehicle with a more or less conventional internal combustion engine.

With that last step, we begin to sound more and more like just a conventional small car. But we don't envision going that far; we are not trying to just build a small conventional car. We really want the design to be tailored to the neighborhood use, cars operating only on local streets and/or in bike path-type situations.

We're not quite sure we know what we mean by "tailored for neighborhood use"—there's lots of room for better imaginations than ours—but here are some thoughts:

- Very easy to hop in and out of;

- Small enough to park casually;

- Light enough to pull off on the grass sometimes;

- Just enough storage for a few sacks of groceries, tennis racquets, or golf clubs;

- Maybe convertible for those beautiful days;

- Cheap and simple: we don't want to pay for the potential for 70 mph when we'll never be over 25 mph.

Might services be extended beyond today's market niches? Some states permit limited operation of golf cars on public streets: the pattern differs from state to state. Thinking seems to have started out as it is reasonable to operate a short distance from the home to the golf course. From that base, operation at low speed on local roads is permitted. The states and local communities seem to be saying, if that is what folk want, OK.

Before proceeding, we present an aside. We want a word to describe a vehicle, a word that gives a vision different from "automobile." NHTSA says golf car and also refers to low speed vehicles (LSVs)—maximum speed 25 mph. Golf course managers say a golf cart has two wheels and is pulled by hand. No matter what others say, we will invent terminology that works for us. We imagine a golf car-like vehicle that is chiefly used for the functions of neighborhood travel and say "neighborhood cars."

The car might also be termed an aid-to-walking vehicle. It occupies the "too far to walk, too close to drive" market niche.

Imagine more

There is no end to this kind of imagining. Indeed, we think of commuter and neighborhood cars as metaphors standing for all the opportunities still out there for vehicles tailored to specific market niches. Such vehicles might improve services and have very favorable impacts on our neighborhoods, on congestion, and on energy consumption [4].

When a family had just one car, it was the general purpose vehicle, and it was nice to be able to transport the whole family. But as we have already noted, times are changing. More and more families own more than one vehicle and can begin to specialize their vehicles for different purposes. The sport utility vehicle, for example, was just a sliver of the car market in the early 1990s, but not so today. And as time passes we will almost surely travel farther down that road of widening the diversity of our road vehicle fleet.

There will be far more variants than the ones we've described here. We even have variants on the ones we've described. For example, we could add a trailer to the commuter car so that the plumber could bring some pipe and the painter his ladders.

Vehicles could adopt smart car technologies, extending perhaps in the future to automated parking, parcel delivery, or pick-up of non-drivers—send the car to pick up Sally after the Little League game.

One service might tie to another. The commuter or neighborhood car might play a station car role—service to or from the train or transit station. With modern communications and payment systems a pool of cars might be available for rental—a commuter might put a card in the first vehicle in line, use it for a time, drop it off somewhere, and be billed for use at a later time.

Need a vehicle for a six-person skiing trip; take a dozen guests to a rodeo; take a date to the opera? Drop in at the neighborhood rental office and rent a vehicle tailored to needs. Perhaps a typical family might have a van in addition to a commuter and/or neighborhood car and rent other vehicles as needed.

Getting started

We only ask that proposals fit present-day markets and travel situations sufficiently to get started and thus have a chance for growth. We need to dissect that wish.

Golf cars and other prototypes for neighborhood-type vehicles have been available for years. But as we've pointed out, we need more than that. And when we add an internal combustion engine, design to conform to NHTSA standards, and exercise the imagination to tailor the design to neighborhood use, we've gone well beyond the golf car. There is a vehicle to build on, but it appears to us that this is a start-from-scratch situation.

For the commuter car there are already vehicles that might serve as starting points—we mentioned the General Motors Lean Machine, and we know of some varieties of motorcycle vehicles that could be thought of as prototypes of commuter cars. Again, there is a vehicle to build from, but an almost start-from-scratch situation.

What is in the way of getting started?

Imagine that a conventional car manufacturer wants to implement a design of either of these vehicles. To keep costs down and assure things will work, the engineer/designer wants to take things out of existing parts inventories (the parts bin, as it is said in the business) and build the car. The use of the parts bin will keep costs below those that would follow if the design and building of new lines of parts were required. But built of old parts, how can something be new, how do we achieve "performing differently."

Suppose the boss says, do as Porsche did, don't build a small conventional car, design something that has a lot of new in it, which the Beetle had. That is not impossible. General Motors did pretty much ignore its parts bin when designing its electric car. (But circumstances for the electric vehicle were special. They seem to involve a mixture of cost-is-not-an-issue, public relations, and manufacturing innovation motives.)

There are still more conversation stoppers. Producers and sellers of automobiles in the United States commit to providing a supply of parts for a new design for over a decade. This is to protect consumers. One would not want to buy a car only to find after a year or so that parts are not available or must be specially made at high costs. This parts requirement also makes the fielding of new products risky to a car manufacturer. The manufacturer might be forced to produce and stock parts for a small market. To avoid this situation, novel vehicles may be rented to users.

Finally, we need to mention manufacturers' liability. The automobiles we know have track records and one knows something about liability—enough to cost out liability and do product and market planning. A new vehicle much like existing vehicles would fit their risk-liability pattern. Something new and different would not have such a track record.

So bringing either of these vehicles into being is not a trivial decision, but neither is it hopeless. It really depends almost entirely on how the market is perceived: designing and producing a truly new vehicle would require lots of confidence that a big market is there. We will discuss the operating environment and the potential impacts of both of these vehicles next; this will begin to give us some feeling for the nature of the possible market.

Operations and impact—the commuter car

The commuter car is envisioned to travel along with automobiles, motorcycles, and trucks on congested freeways as well as on arterial streets and local roads. At first they would appear in mixed traffic streams, but in time we can imagine half-width commuter-like cars operating on narrower lanes, preferably half-width so that two lanes could be carved from an existing, say, 12-ft lane. Provided the demand was there, that would certainly make for much better use of road space. Being shorter than a conventional car matters, but the big gain is from the narrower width (Figure 6.2) [5].

Half-width lanes are easy to imagine, but there would have to be a lot of commuter, half-width cars in the traffic stream to warrant replacing a conventional lane with two narrow lanes (or perhaps two conventional lanes with three narrow lanes). This, of course, gets back to how large the market is.

There is still the sticky problem of a foot in the door, so to speak. Is there a way for the first adopters of commuter cars to achieve cheaper and faster on current streets with current traffic? It would appear that there is. Traffic laws for motorcycles in most states allow their use on HOV lanes and their weaving in and out of traffic while passing in congested

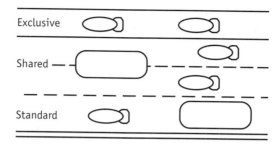

Figure 6.2 Roadway options for commuter vehicles [3].

situations. A commuter car might be picked as a superior product by some persons considering a motorcycle purchase—travel with the radio on and arrive at work or wherever not windblown.

Such a car might well enable new modes of work or social activities because of the ease of reaching dispersed sites. The demand question turns on both the economies the car offers as well as making travel easier and faster, thus making for a better interactive life.

That is not to say that the product would be superior on all dimensions. There will remain those who want to be windblown, just as there will remain those that want a regular car to accommodate a companion at their side.

There is also the issue of safety, concern about crashes between light commuter vehicles and heavier vehicles. The commuter car could be built with crush space and air bags. The physics that says that light vehicles fare not so well against heavier ones, other things equal, is valid, but with increased traffic separation and more nimble vehicle performance and maneuverability, safety may not be an actual problem. But perception is important, and it could be a big problem if the vehicle is perceived as unsafe [6].

We do not see major changes in urban land use as a result of such vehicles. We do see, however, greater economic efficiency; we are reducing by one or two tons the weight that has to be moved around by every user. And given success in gaining a substantial market, we do see much more efficient use of road space as we are able to safely use narrower lanes. Clearly, such half-width vehicles will also increase the capacity of our parking lots.

Operations and impact—the neighborhood car

The neighborhood car will not be quite as narrow as the tandem-seated commuter vehicle, but as we envision it, it will be appreciably smaller than an ordinary automobile or small truck.

Once born, what about growth? It is one thing to image a vehicle in a driveway or garage, on the road with other vehicles is a different matter. It may well be that growth would turn on the present availability and expansion of street spaces on which the vehicles could operate effectively.

It should be feasible to carve space from existing routes to accommodate new kinds of vehicles and their services. Even many older residential neighborhoods have street widths ample enough so that paths for small vehicles could be reserved near curbs. One can imagine neighborhood vehicles operating on bike-like paths in urban and suburban areas and, in those developments, designed around their use [3]. We could also see these neighborhood vehicles sharing their lanes and paths with the small automated delivery vehicles described earlier.

One small neighborhood vehicle displacing a conventional car would more than double the available space for parking or special road lanes. So if the market were to unfold, then feed back would open up spaces for vehicle operations and parking. Garage space? We have seen simple kits on the market for off-driveway golf car parking shelters. Away-from-home trip ends (shopping centers, schools, transit stations) should welcome neighborhood cars and make spaces available because they would increase the number of vehicles that can be parked in a given space.

As small neighborhood cars are increasingly used, perhaps some street space previously used by large vehicles might begin to be allocated to play, visiting, and garden spaces. Conventional cars might be parked out of sight at the edges of neighborhoods, and neighborhood cars might be used to access remote parking areas.

Cheaper? Yes. Faster? Not when compared to ordinary cars on ordinary roads, but yes when compared to walking. Better? For special classes of drivers, yes. Less intrusive vehicle operation might enhance neighborhood quality and ambiance. Perhaps neighborhood vehicles might produce better neighborhoods.

So the first round of effects might be cheaper and faster-than-walking transportation for previously house bound grandmother and young Sam as well as for folk already driving automobiles. The second round might be better neighborhoods, quieter with more space available for other-than-road uses. In particular, old neighborhoods where the conventional car does not fit so well might welcome reduced congestion and improved neighborhood quality.

We think that the driver in the neighborhood car market is a better neighborhood package—less invasive full-size automobiles, less vehicle noise, and street spaces available for play and socializing.

What about large vehicle access to neighborhoods where street spaces have been configured for small vehicles. We are thinking of emergency, construction, moving, and other large vehicles. We think that with a little imagination in design, such vehicles could be accommodated without doing too much damage to neighborhood ambiance.

Headwinds

These are not technically difficult items, and the idea certainly isn't new. Why hasn't it already happened? Some people blame it on "market failure."

We admit to being skeptical about blaming "market failure" for the absence of the kind of cars we've conjectured here. There are usually good reasons that such new ventures haven't happened already. But there are also often reasons that just follow from what we might call habit patterns in thinking.

For example, a thoughtful representative of a highway agency once said to one of the authors, "We (folk in the agency) see no market for small cars." That was after a presentation in which the author unsuccessfully argued that standards blocked innovative restructuring of road spaces and its uses.

In hindsight, that was a highly predictable conclusion. It may be that there is not much revealed demand for neighborhood and commuter type vehicles because there is a limited number of streets suitable for the unique operational capabilities of such vehicles. It is the classic chicken and egg problem—no suitable road spaces, no vehicles demanding them; no vehicles, no need to produce suitable road spaces. Also public works agency professionals are proud of their products. When there is a no-demand excuse, it is unimaginable that an alternative product might be superior.

Consider another example. Almost all of the proposals for new things that we have seen assume that present day formats for providing services will continue; they ask for actions that fit existing formats. For example, hybrid power trains (battery and electric motor plus small carbon-fueled engine and alternator for power generation) are targeted for existing small- to medium-sized automobiles—the hybrid is to have similar carrying capacity, driving range, and acceleration. Electric vehicles using batteries also seek to reproduce the performance of existing vehicles. Now even we agree: this happens because it makes sense—there is enough risk already; why try to introduce something else new at the same time if it isn't necessary.

So we do not knock that approach—things must fit existing situations; there is much less risk that way. Producing a similar but improved product is good business strategy.

Yet here we are trying to turn the "fit situations" formula on its head. We are aiming at existing market niches, then predicate growth being pulled by the activities enabled by new services, the uncovering of new market niches.

We repeat: we view neighborhood and commuter cars in a metaphoric way—they stand for new combinations of vehicles, facilities, and uses providing new services and new opportunities. In this context, there is plenty of room for wider imaginations than ours, and they should be encouraged. Our cities should have something for everyone.

References

[1] Ministry of Transport, *Cars For Cities*, London: Her Majesty's Stationery Office, 1967.

[2] Riley, Robert Q., *Alternative Cars in the 21st Century: A New Personal Transportation Paradigm*, Warrendale, PA: Society of Automotive Engineers, 1994.

[3] Garrison, William L., *Studies of Road Infrastructure Requirements for Small Innovative Vehicles*, Berkeley, CA: University of California, Institute of Transportation Studies, UCB-ITS-PRR-93-16, 1993.

[4] Sperling, Daniel, *Future Drive: Electric Vehicles and Sustainable Transportation*, Washington, DC: Island Press, 1995.

[5] Herman, R., T. Lam, and R. Rothery, "An Experiment on Car Size Effects in Traffic," *Traffic Engineering and Control*, Vol. 15, June 1973, pp. 90–98.

[6] National Highway Traffic Administration, *Small Car Safety in the 1980s*, Washington, DC: U.S. Department of Transportation, 1980.

Part II

The City and Transportation

7

Transportation and the Evolution of the City

Born before recorded history, the city is still a work in process.

The birth of the city was quite possibly the first great watershed of human history. "Birth" is probably the wrong word, for it implies an event, and the city is a very vaguely defined region in the continuum of evolution from the family to the clan to the village to the town to some higher level of organization and size that we label "city."

But no matter. The city provided humankind its first opportunity to behave collectively on a large scale: to achieve the security of numbers; to satisfy the social and religious nature of mankind; to let many minds interact; to de facto experiment with various forms of political, economic, and social organization; and to amplify the exploitation of specialization and the division of labor. The city was the start toward civilization as we know it.

We take for granted that cities need transportation and give that fact little more thought than a fish gives water. Many of us instinctively take an egocentric view of that need and say we need the transportation that lets us live happily in the suburbs and the transportation that keeps the supermarket full of groceries, for example.

Unless we stop to think very hard, we fail to appreciate the key role this transportation has had in shaping our cities, in influencing their locations, and in facilitating their growth and interactions. Of more interest to us now is the influence it will continue to have in the future. Our purpose here is to illuminate this influence, this process, and we start by drawing upon the past.

The first requisite for a city is a reliable supply of food and water. This is clearly fundamental, and it's no coincidence that the first agriculture and the earliest cities appeared on the fertile plains off large rivers: the Indus, the Tigris–Euphrates valley, and the Nile. The river was both the water supply and the primary transportation corridor.

The only ground transport was walking; goods were carried on the backs of animals, including Homo Sapiens. Real roads were almost nonexistent, both hard to build and hard to maintain, and this lack kept extensive wheeled transport a long, long way in the future.

Improving transport, mostly by water, let cities outgrow the agricultural capability of their own immediate fields and import foodstuffs from more distant regions. Better transportation let food supply slowly recede as the limiting constraint on city size. Ancient Babylon lived by river transport; and both Athens and Rome depended partially on Egyptian grains, carried over water.

Not only was good external transportation a precondition for growth for supply reasons, but it was a prerequisite for general commerce and interaction between cities. Transportation enabled the long transition from a world of city-states to a world of nations.

Transport by water enjoyed two sources of propulsive power that was rarely practical for land transport: wind and, in special situations, river currents. There is evidence that the sail was used by both the ancient Sumerians and Egyptians at least five thousand years ago. Except for those rare situations in which the whole trip is downhill—like the trip from some English coal mines to the watercourse—land transport depended entirely on muscle power until about 200 years ago.

It took thousands of years for the wheeled vehicle to make appreciable inroads into feet as the primary mode of personal transportation and
goods movement. As we noted earlier, we overrate the invention of the
wheel because we forget that it takes a suitable road to be of much use.
For thousands of years the potter's wheel was probably more important
to civilization than the wagon wheel.

Early roads were for walking; the Persians operated a "pony express"
for fast transport and forbade wheels because they tore up the roads.
Although some wheeled vehicles were used, most Roman roads were
designed for walking traffic.

Thus, most transportation between cities was slow in 3000 BC, and it
was still slow in 1800 AD, not over 3 mph on average.

And it was also slow within cities; proximity had to substitute
for mobility. Cities were dense because the low velocity of most
transportation demanded that homes, markets, and places of production—jobs—be close together. If they were not, then transportation
between all of the places that an individual needed to go in the course of
daily living ate up an inordinate share of the day's hours. A rule of thumb
that appears to have remained valid over the ages is that few people will
spend much more than an hour or so in routine, daily transport.

The need for walls, and the desirability of being inside them, also was
an incentive for crowding.

The description of the city of medieval Europe sounds a lot like
that of the Indus river city of Mohenjo–Daro in the fourth millennia
BC—except perhaps the sanitary arrangements in Mohenjo–Daro were
more advanced. The premium on land space and the consequent crowding of both habitat and activity sites squeezed the land allocated to streets.
Most were really alleys just wide enough to accommodate walking by
both people and animals. Since some streets also served as public space
and as markets, those leading from the gates in the walls to the center
of town were a bit larger, but still cramped by modern, automobile-
city standards. We would not be favorably impressed by the living
conditions.

So all those quaint European cities we love to visit, with their narrow
streets defined by buildings with overhanging upper floors, an open
square that provides enough space for markets, didn't get that way just
because people cherished togetherness.

Our central point in all this is that for thousands of years there were no real fundamental changes in the spatial arrangement of cities because there were no fundamental changes in the transportation technologies that would permit or enable those changes. Gun powder and the cannon finally made the wall obsolete, and wheeled vehicles coupled with growing foot and animal traffic slowly encouraged widening the major streets.

Little changed the basic demand for proximity until well into the nineteenth century. Cities were still dense because shops and stores still had to be close to customers and factories and other commercial enterprises near their workers. And no one who walked to work in the city could live anywhere else. While there were many cities that were much larger than in earlier millennia, walking distance was still a constraint on local spatial arrangements: the functioning city still had to have everything close to everything.

Then less than 200 years ago, we finally began to get the new forms of transportation that ushered in a new dynamism in the spatial organization of our cities. As we already noted, these changes in transportation were part and parcel of the wave of technical evolution we characterize as the industrial revolution. These changes let the city reorganize itself.

The birth of the modern city

The precursor step in the larger cities was the horse-drawn streetcar. Then a few steam-powered cable cars came along that were quickly followed by the electric trolley, which had become common by the beginning of the twentieth century.

The trolley—at least along the trolley lines—roughly doubled the average speed of urban travel, from the 3 mph for walking to maybe 7 mph. With the same investment in commuting time, people choosing to live along the trolley line could have their homes twice as far away from their work as when they had to depend on walking. Thus, the trolley gave birth to the first suburbs.

Cities also grew upward as well as outward: cast iron and then structural steel combined with the elevator to make upper stories possible and practical. More stories on buildings meant more activity in the downtowns; and with the higher density of activities came continued, and probably worsened, crowding. There were enough horse-drawn carriages and carts and wagons (and people) filling the streets to give a healthy foretaste of the congestion that we think of today as belonging in the exclusive domain of the automobile and the truck. Even so, many people, perhaps most people, still walked.

Then came the automobile. In the 1920s the United States went from one car for about every 35 people to one car for every 5. As one would expect, the 1920s was also a decade of road building and road improvement.

While the car removed the constraint that kept folks downtown or within walking distance of streetcar lines, the depression of the 1930s made it unaffordable for many to move farther out. The real automobile-based explosion of suburbs in the United States didn't start until after World War II, when the other forces encouraging suburban growth came into being: the formation of lots of new families, increased real incomes, and government programs aiding the financing of housing.

The impact of the automobile (and the truck) far exceeded that of the step from walking to the trolley. Whereas the trolley was roughly twice as fast as walking, the average speeds offered by the car were more like seven times faster—over three times faster than the trolley. Tripling the speed meant that commuting distances could also triple with the same investment in travel time. Further, there were far more roads along which suburban expansion could occur than there were trolley lines—and greater incentive to add more.

Today, with our freeways and interstates and still higher average speeds, we have "suburbs" 20, 30, and more miles from the original city center. These new settlement patterns have produced a steady decline in the population density of our urban regions. Boris Pushkarev and Jeffrey Zupan in their book *Public Transportation and Land Use Policy* note that the range of densities of settlements shaped by the automobile tend to be only one-tenth of those prevalent historically.

At the same time, the truck freed the factory to move out to cheaper land and to improve productivity by taking advantage of the increased space to reorganize itself and its production processes.

The net effect of all this was that the "city" reshaped itself into a largely low-density "metropolitan area": an old dense core surrounded by low-density development. Our new cities, like Tucson, have grown up around the car and the truck and are almost all low density.

Better transportation decreases the need for keeping large inventories of goods in stores because it becomes feasible to resupply faster. As direct factory to store becomes increasingly feasible, the need for inventory in intermediate warehousing decreases. We are seeing more and more direct factory-to-customer deliveries. The whole pattern of manufacturing and distribution has changed and continues to change.

The truck and the car also make it easier to carry out some of the support functions of running a city: police, fire protection, trash collection, medical support, and the general response to emergency situations.

And better intercity transportation extended the market reach of individual cities, permitting higher output from their industries. The resulting economies of scale and opportunities to organize production in new ways resulted in still further improvement to productivity.

This evolution continues. Store keepers' motivation has not completely changed over the years—they still want to be near their customers. So they have followed them to the suburbs. Only now "near" doesn't mean walking distance, it means easy driving distance with plenty of parking. The mall, an aggregation of shops and stores surrounded by a big parking lot, was born.

Look where we've come. Even 100 years ago most cities were still dominated by a single nucleus where people lived, worked, shopped, and died—a description that differs little from the city of 5000 years ago. First the trolley and then the automobile have been the primary agents of change of that age-old pattern. With the advent of the car and the truck, the city has evolved into multiple commercial and shopping nuclei, all surrounded by acres and acres of largely single family homes, punctuated here and there with a few higher density living complexes and in other places with commercial and manufacturing enclaves.

The quality of life

It is our belief that that the nebulous notion of quality of life hinges in large part on the richness of options for how we shall live. In almost every aspect of our lives we have enjoyed the ubiquitous trend toward broadening choice: in our homes and their location, in what we eat, in nearly every type of product or service, in where we shop, and in how we are entertained—in almost all aspects of living. We see nothing wrong with continuing to encourage this trend, with broadening our options still further.

From the transportation perspective this involves not only alleviating our major problems but encouraging the innovation that has given us the diversity and variegations in transportation that have contributed to our choices today. Almost every day a new market niche of some kind or another is identified and served. Our interest here is to throw some light on both additional possibilities as well as the processes through which this happens.

It is not our thesis that transportation shapes our cities. But it does help determine our options by either enabling and constraining the way the city shapes itself. The reshaping is the result of the citizenry exercising their preferences in how to carry out the functions of living—producing and exchanging goods and services, raising children, playing, worshipping, and studying.

These preferences and choices are motivated by all kinds of personal reasons. If transportation is poor, more of these preferences will be made infeasible. Both historical experience and logic suggest that transportation that has high costs or is unreliable, slow, or difficult to use will result in both lower economic productivity and constrained social activities. As transportation improves these constraints drop away, more preferences are realized, and the city becomes more efficient economically and, on average, can reasonably be expected to be a more satisfying place to live.

Transportation does not cause these city-shaping motivations; it is a support function that enables or constrains their realization.

It is sometimes argued that the suburbs have expanded not so much because they were preferred by the majority but because misallocation of costs have made them cheaper than their true cost. There's almost no doubt that economic factors were a powerful part of the calculus, and sometimes costs are misallocated, distorting such calculations. But the movement has been too overwhelming to believe there is not a strong preference by the majority of families for suburban living. The car and the trolley didn't make people move out of the city, it let them move.

If people did not want to live in suburbs, the car would not make them do so.

While we are highly skeptical of our collective ability to plan cities, we recognize that some constraints on pure market forces are probably needed. We should strive for public sector actors insightful enough to supply these constraints without closing off preferred options. Within these constraints, it appears to us that the most desirable course of action is to provide a climate that encourages a flexible and versatile transportation system, a system that permits the realization of as many preferences of the citizenry as we know how.

Whither now?

There is no argument that the picture is complex, replete with tradeoffs: more living space per person versus the preservation of natural and rural environments, development here intensifying traffic there, balkanized political entities dealing with area issues. Some of these problems are seemingly insoluble and a source of great frustration to those who are trying to solve them.

Viewed in the historical perspective, it's not too surprising that we find ourselves in this spot now and then: in the last 100 years or so we have had to cope with a new world, a world that is changing more rapidly than at any point in mankind's history. We have to remember that we humans have muddled through problems before and are not completely bereft of ideas about how we might muddle through again.

In the next chapter we will try to sort out some of these problems, trends, and influences and hopefully develop a clearer and more rational picture of just what broad paths into the future seem most sensible—the preferred directions for muddling, so to speak.

Following that, we will lay out an example of an alternative approach to transportation that might enable a different kind of urban spatial organization. It is an example, illustrating that there may be more to the future than cars and trucks and transit as we know it. It turns out that this scheme suggested another application, which we briefly outline in Chapter 8.

Now completely new schemes might be very nice, and they just might happen. But then again, they might not. And there is more to the picture than just laying out the broad roles for the car and the truck and for transit as we know it. That is the top level, and it's important that we get it right. But the richness is in the details.

So in the closing chapters of this section we first examine some of the current features and trends of our variegated cities, which reflect an almost continuous evolution of transportation as it responds to the preferences and needs of their populace. We then focus on the generic processes of innovation, the processes through which the diversity and variegations in urban transportation have evolved, to identify the kinds of generic actions that might encourage the realization of a preference-responsive city in the future.

We have set before us a daunting agenda. Woody Allen remarked that time was invented to avoid the confusion created by everything happening at once; we use chapters to avoid the confusion created by trying to say everything at once.

Selected bibliography

Ausubel, J. H. and R. Herman (eds.), *Cities and Their Vital Systems: Infrastructure Past and Present*, National Academy of Engineering, National Academy Press, 1988.

Konvitz, Joseph W., *The Urban Millennium: The City Building Process from the Early Middle Ages to the Present*, Southern Illinois Univ. Press, 1985.

Mumford, Lewis, *The City in History: Its Origins, Its Transformation, and Its Prospects*, New York and London: Harcourt Brace Jovanovich, 1983.

Stuart, G. E. and J. Tapper (eds.), *Peoples and Places of the Past*, Washington DC: National Geographic Society, 1983.

Pushkarev, Boris, and Jeffery Zupan, *Public Transportation and Land Use Policy*, Bloomington and London: Indiana University Press, 1977.

Saalman, Howard, *Medieval Cities*, New York, George Brazillar, 1968.

Zahavi, Y., M. J. Beckmann, and T. F. Golob, *The "UMOT"/Urban Interactions*, U.S. Dept. of Transportation, DOT RSPA DPB-10/7, 1981.

8

The Modern Dilemma: What to Do?

The first step in solving a puzzle is to turn all the pieces right side up.

The automobile and the truck have broadened our options in the way we live and work beyond anything we could have imagined 100 years ago.

The automobile and the truck have also brought us some very confounding problems and some very tough choices.

The spectacular growth and enormous popularity of cars and trucks in comparison to other transport modes has clearly had a downside. More and more of our cities suffer from congestion. We also hear about energy and environmental issues, urban sprawl, and how traffic hurts neighborhood serenity [1].

These problems are real, and they are there for all to see. It is not too surprising that in many people's eyes the car has become the villain [2], not the marvelous instrument that has brought us a richer life.

These same people view transit as the potential savior, even though people voting with their money have overwhelmingly chosen the car. That choice is easily understood: transit lacks the speed of the car, the on-demand readiness of the car, the freedom of route choice of the car, the versatility of the car, the personal privacy of the car, and other attributes social scientists emphasize [3].

When one reflects on these differences, it is easy to understand why the car dominates urban transportation, providing something over 90% of all urban passenger miles. This dominance has been growing over time.

In spite of increases in the urban population, transit's share of passenger trips went from about 20% in 1955 to about 5% in 1985 and continues to decrease. We have spent about $385 billion in subsidies trying to reverse the trend; and these expenditures, which have risen four times faster than inflation, have been increasingly ineffective [4].

And the automobile has let the city reshape itself so as to make it still more difficult for transit services as we know them to be competitive.

We think the prospects for reversing this trend, for having people willingly abandon their cars and take to public transportation on a widespread basis, are exceedingly slim.

In the face of this steady decline in transit ridership, it is surprising that transit appears to enjoy more public support than improvements to the road system [5]. But we agree with those that think many people support expenditures on transit because they hope it will get the other drivers off the road. They think they are supporting a cure for congestion.

We also suspect that pro-transit sentiment partly reflects nostalgia, analogous to the yearning for the return of the steam locomotive. It's an understandable sentiment: in our minds we can still see Judy Garland singing "Clang, clang, clang went the trolley" in a charming if perhaps partly fictional picture of our past. But rail transit was truly worth singing about in those days: it was a great leap forward over its then-competition, either the horse-drawn conveyance or one's own two feet. That era still competes in our minds with today's reality. But uncritical nostalgia doesn't serve us well today [6, 7].

Planners often put forward the notion of self-contained neighborhoods: closely spaced work, residencies, shopping, and recreation—there are many variants and many designs [8]. Planners tell us we should deplore "sprawl" and live in these more "efficient" communities. Part of their motivation is concern about paving over all our open and natural spaces with homes. But in many cases the dominant motivation is providing a high enough concentration of potential riders to get more people onto transit.

With a few exceptions, the public seems to have rejected these ideas. What many do appear to want is a small town "feel," but they want that "feel" to go with a house in the suburbs with its own backyard [9].

There is almost no question that low-density suburbs have costs that go beyond just the cost of housing and land. About 25 years ago the Real Estate Research Corporation did a paper-and-pencil (no reference to real data) study of the costs of sprawl and found it costly on a number of dimensions, including pollution and energy consumption [10]. A more recent data collection effort reached similar conclusions [11]. These works are widely known among professionals, and their methodological and measurement problems are well recognized. But their major deficiency is that they are largely silent on demand questions—given these costs, are people willing to pay them? Clearly it depends on their magnitude relative to incomes; in Europe, where consumption is taxed much more heavily than in the United States, suburbanization has grown more slowly than here [12]. But all over the world, and with very wide variations in the level of such costs borne directly by the suburbanite, as affluence rises, so does the growth in suburbs [13].

The complexity of the problems and the frustration stemming from the lack of easy solutions seem to have led many of our planners and policy makers into promulgating prescriptions that often seem to border on desperation: hope springs eternal when no other "cures" are perceived. The most common prescription is not doable in any practical sense: severely restrict the car and expand transit everywhere. For good reason, people like their cars; the most prominent goal in the life of most teenagers is still their driver's license. And, as we have repeatedly reiterated, transit is a very inefficient option in low-density suburbs. The major concern is that we have let traffic get out of hand.

Bicycles are part of the prescription sometimes. The pictures show young, healthy, happy cyclists riding on beautifully landscaped bike paths on bright sunny days, their laptops apparently safely stowed in their backpacks.

Do we really want to turn back the clock, to reverse time's arrow? It appears that the vast majority of the population do not. And we couldn't, even if we wanted to.

But we are paying a high price by trying. For whatever the reasons that perceptions and prescriptions are distorted, the result is a climate of opinion that tilts both planners and decision makers into trying to force transit into niches where it doesn't belong, to reshape the city to its 1920 configuration. Too much effort is going into imposing already known and outdated pseudosolutions, and not nearly enough effort is going into discovery and experimentation with new ways of making our road mobility system better and more capable of serving our future.

The reality is that, for better or for worse, the use of automobiles and trucks show little sign of diminishing, nor does the growth in household ownership and the demand for larger homes and suburban living. Our primary focus should be on accepting these trends and taking those actions to accommodate them as well as we can, recognizing that nearly every action we take involves tradeoffs.

One of these trades involves the long-run allocation of land and the preservation of the natural or rural environment. We do not even remotely suggest that we have an answer as to how that should be done, but we do suggest that letting our road system choke with congestion as a means of allocation will not result in an improved quality of life.

Taming congestion has to be high on our list. We are only slightly moved by the arguments often put forth that only a small percentage of our road space is chronically congested: apparently it's sheer coincidence that congestion occurs just where the most people want to be. We have already concluded that if we are to significantly alleviate this problem, we have to take some combination of the steps we discussed in Section 1: improve our traffic management system to use our existing road system more efficiently, including better control over usage level, and to either add more concrete or, sooner or later, impose pricing constraints on access. The last is a "solution" we would rather forego but probably cannot do so completely.

In the remainder of this chapter we attempt to illuminate what we consider to be the primary realities that influence the directions we might take. Our need is to not only try to alleviate the major problems with which we live now but to anticipate and accommodate the future.

Here we will paint with a fairly large brush, sticking mostly to the roles and outlooks for automobile and transit services. In later chapters we will pick up some of the smaller brushes. We will also defer energy and environmental issues until later.

Relative roles

No, we can't build enough roads to use the car everywhere—it just isn't practical. Just as buses and suburban rail have their niches, cars also have their niche, albeit a pretty large one. We will look at what drives these niches.

The primary reason the car can't be used with impunity everywhere is that it takes up too much street space. The vehicle itself takes up 15 ft or so of lane. To that we have to add the space that drivers maintain between vehicles, roughly 20 ft to 30 ft at the low speeds of downtown traffic. Together this gives something like 40 ft of lane length needed for every car, a car that usually carries just a single occupant.

This need for lots of street space per car is not a problem in low- and moderate-density suburbs because the number of trips the streets have to accommodate is low. But it is a problem in denser developments and in the central core of all but our newest cities: the trips in and out of the area and the trips made within the area are just too numerous to be served exclusively by the space-hungry automobile. Remember that we also have to consider space for parking and that used by pedestrians. Put all these together and dependence on the car alone for transportation in higher activity density areas is just not practical. If these areas are to be properly served, some type of transportation that is more space efficient is desired.

Now compare in Figure 8.1 the automobile's requirements for street space with those of a downtown bus, maybe 40-ft-long with the same 20 ft to 30 ft of safe spacing between vehicles. Even half-full, this is only 3 ft or so of lane per passenger; full, it's only 1½ ft of lane per rider. As a rule of thumb, we wouldn't be far off to say that on average a downtown bus is 20 times more efficient in its use of street space than a car.

This is a huge advantage for bus transit in our high-density downtowns where there are a high premium on street space and lots of trips to keep the seats filled.

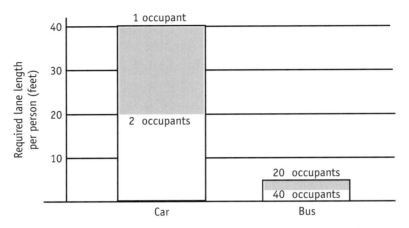

Figure 8.1   The length of lane needed to accommodate one occupant.

Light rail only takes up street space at intersections—it has its own "street." The surface space required by underground trains is that needed for passengers to enter and leave the system. So rail systems win the street space–efficiency contest hands down. But buses can run up and down every street and so can potentially provide much more comprehensive service than rail systems whose networks are much more sparse. Therefore, buses are widely used even in those cities that have rail systems.

Obviously cars can also be put underground, thus reducing their impact on the surface, but they still require lots of lane space; it's an option, but an expensive one.

An additional argument for transit often put forward is that we need it for those of us who do not have a car or are disadvantaged by residential location, limited employable skills, or income. Census data report that poverty households in central cities are relatively dependent on transit; and each of us knows of other needs, such as those of single car households where more than one person is employed.

But the argument for increased transit services seems weak. It asks that transit be used to solve or ameliorate welfare, education, and discrimination problems that ought to be managed in other ways. For the unemployed, access to jobs matters, but education and training matter even more [14].

Where improved services have a role, services could be provided for some segments of the population by providing properly equipped cars or vans, including drivers. Already, taxi-like jitney services are serving these markets [15]. In the future, the increased automation of car-like services should broaden the availability of services tailored to needs.

Service in the suburbs—thin, diffuse travel patterns

This country began to move away from the high-density urban structure compatible with transit in the 1920s when automobile ownership took its major jump. And except for a blip during the rationing days of World War II, suburbs have since been in a steady expansion and transit has been in a steady decline.

We already know the problem; we discussed it in Chapter 4. In low-density areas, which is almost all suburbs, there just aren't enough trips being made to provide reasonable passenger loads on buses. In the suburbs people move in small numbers to diverse destinations and vehicles designed to carry small batches, like cars, work best.

Transit is designed to carry people in large batches, and it is just not possible to generate large batch movements in low-density areas. The practical cap on expanding fixed-route bus service is hit very quickly in a low-density urban suburb. And while full buses are both energy and space efficient on a per-passenger-mile basis, empty buses are not; the car is more superior to a lightly loaded bus on a per-passenger-mile basis.

The economics of rail transit demand even higher passenger flows than buses. So rail transit is also not appropriate where travel patterns are diffuse: too few trips start and end near a transit station to permit heavy ridership. If travel distances are long, people are willing to put up with the hassle of park-and-ride or kiss-and-ride, so rail transit is most often seen serving the long-distance commutes into large cities and commercial districts.

In sum, where travel is heavy all the way from origin to destination, there can be an important role for both bus and rail transit, but there is no real competition for the car in low-density areas. We conclude that the optimum low-density suburb mode of transport from the viewpoint of both the cost and convenience is the automobile.

But once in the car

Unfortunately for transit usage, most people live in suburbs; and since most trips either start from home or end at home, most trips either start or end in a car. And once in their cars, most people don't want to get out of them until they get where they are going.

If people start downtown in their car—and most have no remotely competitive alternative—they drive their car downtown in spite of the traffic problems produced. And the downtown merchants and businesses are very much against telling them that they can't; they want to make it easier, not harder, for the customer to reach them.

The car travel generated in the suburbs gets channeled into large flows on freeways or large arterials, and if suburban areas continue to expand, the freeway traffic will do likewise to the degree such travel is not diverted into other patterns by the concomitant movement of businesses to new locations.

The popularity of the suburban shopping mall and the decline of downtown shopping stem largely from the relative ease of access and parking for automobiles provided by the former.

There are, of course, exceptions. As we noted, where trips are long and aimed at a central downtown or other high-density enclave, people do use commuter rail and express bus alternatives. Here is where park-and-ride (or kiss-and-ride) enters the picture.

But by and large, people who want to go downtown from the suburbs start the trip in their car, and just because buses and undergrounds and rail transits fit better downtown, they are not always willing to park the car and board one of these alternatives for the rest of the journey. Many do, of course, but downtown streets are still full of cars, usually looking for a place to park.

Now one possibility is to restrict cars from these downtowns by fiat; this is done in a few foreign cities with more tractable citizenry than is typical of this country. But if the purpose is to increase the economic health of the downtown merchants and other businesses, constraining access may not be such a good idea.

Where does this leave us?

First, the car as we know it can't provide adequate transportation in downtowns and other areas of high-activity density. These areas require an alternate form of transportation. Today that is the bus and perhaps rail transit, either underground or on the surface. Variants like dial-a-ride work in moderate density situations. Maybe some form of personal rapid transit (PRT) systems—small, four- to eight- passenger "cars" running under computer control on a fairly dense network of small tracks—have a role here.

Second, these forms of transport that suit high-density development are unsatisfactory in trying to serve lower density dwelling or activity areas. Here the car is optimum. We spend lots of money sending transit into suburbs to poorly serve a very thin market.

Third, it hasn't worked well to try to force people to transfer from cars to transit or vice versa, so each of the modes can be restricted to where it works best. Most people want to use their cars everywhere.

What, then, to do?

About our downtowns?

We have no magic bullet. We suggest two broad courses of action. The first is to make it as easy as possible for people to get around within a high-density area without their cars; this makes it easier to persuade people to park on the periphery. Ordinary bus service doesn't seem to be able to make much headway along these lines, so something better is preferred. Personal rapid transit? Still another possibility is described in Chapter 10.

The second course of action is to make the center as tolerant of vehicular traffic as possible. The ability of downtowns to cope with traffic can be improved by such common-sense steps as separating vehicular and pedestrian traffic, ensuring plenty of off-street parking, and providing locations where people can be let out of cars or picked up without the car having to interfere with traffic in the process. The same considerations apply to trucks and other pick-up and delivery vehicles.

So we can either find ways to better accommodate the car everywhere or make it more attractive to leave the car outside areas where it doesn't fit.

Emulate Europe?

The public transportation system in Europe is often held up as a goal for this country. Their public transportation system is much better than ours; and the argument is made that if we were to improve ours to the same point, then transit would make a comeback in this country. The fact is that Europe is following us, not leading us. The level of car ownership per capita that the United States reached in 1929 was not reached in Europe, which had been devastated by World War II until 1970 [16]. So the Europeans have had some 40 fewer years for the car to both erode transit through the competition it offers and to reshape their cities to its further detriment.

For better or for worse, Europe is clearly moving along a development path similar to that of the United States. For instance, Paris has lost about one-fifth of its population to low-density suburbs and transit ridership is declining [17].

Transit in perspective

There are many new technologies now available and rapidly coming into being that can improve the quality of service transit can offer; as noted, the intelligent transportation system program is a prime mover in this area. But we can't let ourselves get too beguiled by new hardware—shiny cars or automated this or that only improve service at the margin. The fact is that the functional characteristics of rail transit today are little different from those of the first streetcar that poked its head into the suburbs over 100 years ago. They both follow a fixed route on a hopefully fixed schedule and provide service by stopping at prespecified points to pick up and discharge passengers.

But there are two big differences between then and now. Then, a lot more people lived near each stop at the bedroom end of the trip than today. Today, with our much lower density suburbs, far fewer people would find it convenient to ride. Second, the alternative transportation of a century ago—feet or horse-drawn conveyances—was much less attractive than today's automobile, so the likelihood of transit use was much higher back then. New hardware can't change these two fundamental facts.

In spite of its modern shortcomings, transit today plays a very important role in the life of most of our larger cities, and we applaud and encourage efforts to make it even better in these roles. In nearly every city of any size, transit is important during commuting hours. There are many places where transit makes a major difference in our urban quality of life; the Washington DC Metro is an example. But even here usage has been decreasing in recent years.

But while transit may be slowly declining, we will live with it and benefit by it for many, many years and it would be very shortsighted not to make it as good as we reasonably know how—within reasonable economic bounds. We don't hold back on investment in a home just because we know that we won't live there forever. What we do criticize are efforts to make our world revert to 1920 and to let that subliminal goal lead us into pushing transit where it is not appropriate, making investments that are not commensurate with the benefits they will bring.

Overall, we would encourage the further expansion of public transportation if the changing pattern of urban infrastructure—the built environment—provides valid opportunities for its sensible application. We advocate, however, that to the degree possible we let the people decide how they want to live, not the planners.

The road system

It was not by chance that we devoted the first section of this book to the road vehicle system, specifically, the problem of congestion and possibilities for the future that included both new vehicle designs and the prospects for the full automation of vehicles.

The road and freeway system is by far the dominant transportation of our age, and the quality of urban mobility into the long-range future will depend almost entirely on its quality and performance.

This last sentence bears repeating: *the quality of urban mobility into the long-range future will depend almost entirely on the quality and performance of our road and freeway system.*

We forget this central point when so much of our energies are promoting plans and schemes aimed at carrying us back into a irretrievable past. Those energies could be much more fruitfully focused on finding ways to improve our future.

It strikes us that a sensible first step toward improvement of transportation in almost every reasonably large city is a frontal attack on the congestion problem.

In addition, there are a myriad of other ways to improve the safety and convenience of personal transport, such as better traffic management, and law enforcement, and more useful traveler information. Specialists can point to many others. We will discuss air quality in a latter chapter, but that, too, can be improved to the point that it is a secondary concern.

These actions should be accompanied by a program of road infrastructure improvement and expansion as necessary to cope with the anticipated growth in vehicular traffic in the future. This sentiment is considered both heretical and recalcitrant in many circles, but we suspect it is something most everyone knows but is reluctant to accept.

Innovation

Change is not the enemy. Fostering a climate that encourages initiative and innovation in new approaches to supplying transportation, to filling niche markets, and to better tailoring transport to the function it serves could, in the long run, be as important as the various actions we've already noted. While it is fundamental that overarching, "big brush" problems like congestion need to be alleviated, much of the richness of our lives derives from the details, the availability of many kinds of tailored transportation options and the variety of services that derive therefrom.

We already gave some examples of alternative kinds of cars in the prior section, and we feel sure that there are variants of today's transit systems that could better adapt them to tomorrow's circumstances. An example from the past is dial-a-ride. As it became increasingly obvious that fixed-route bus service was very inefficient in serving lower levels of ridership, some enterprising souls conceived the idea of having riders call in their need for transport and the minibuses respond directly to these calls—a kind of hybrid between the bus and the taxi. It has turned out to be very useful in moderate ridership environments. There also may be merit in the PRT concept that is being pursued in so many places [18].

There are now commonplace many kinds of mobile services, each aimed by its owner at making their own living, but serving, as Adam Smith told us they would, to make our lives better. We think this subject of innovation, of filling market niches, is very important to the richness and quality of urban life and have devoted several later chapters to it.

Next

In the next two chapters we will hypothesize several possibilities for future evolution in patterns of living and working. We will discuss the notion of a variegated city, of cities responding to the preferences of their populace, and the role of innovation in their evolution. The last chapters address the kinds of generic actions that might encourage the realization of the citizen-responsive city.

References

[1] Owen, W., *Cities in the Motor Age*, New York, NY: The Viking Press, 1959.

[2] Jerome, J., *The Death of the Automobile*, New York, NY: W. W. Norton, 1972.

[3] Sandqvist, K., *The Appeal of Automobiles—Human Desires and the Proliferation of Cars*, Report No. 1997:21, Stockholm: The Swedish Transport and Communications Research Board, 1997.

[4] *Transit Fact Book*, The Wendell Cox Consultancy: www.publicpurpose.com, 1999.

[5] Gomez-Ibanez, J. A., "Big City Transit Ridership, Deficits, and Policies: Avoiding Reality in Boston," *Journal of the American Planning Association*, Vol. 62, 1996, pp. 30–50.

[6] Richmond, J., *Transport of Delight: The Mythical Conception of Transit in Los Angeles*, Akron, OH: The University of Akron Press, forthcoming.

[7] Richmond, J., *The Political Rebirth of Rail Transit*, Cambridge, MA: Taubman Center for State and Local Government, Harvard University, 1999.

[8] Dantzig, George B., and Thomas L. Saaty, *Compact City*, San Fransisco, W. H. Freeman, 1973.

[9] "Tennis Anyone? Study Reveals What's on Home Buyers' Wish List," *The San Diego Union-Tribune,* San Diego, June 6, 1999.

[10] Real Estate Research Corporation, "The Costs of Sprawl," *Land Costs Analysis*, Washington, D.C.: Council on Environmental Quality, Department of Housing and Urban Development, and Environmental Protection Agency, 1974, chap. V.

[11] Newman, P. and J. Kenworthy, *Cities and Auto Dependency: An International Sourcebook*, Aldershot: Gower Technical, 1989.

[12] Nivola, P., *Laws of the Landscape: How Policies Shape Cites in Europe and America*, Washington DC, Brookings Institution Press, 1999.

[13] Pucher, J. and C. Lefevre, *The Urban Transportation Crisis in Europe and North America*, London, MacMillan Press Ltd., 1996.

[14] O'Regan, K. M., and J. M. Quigley, "Accessibility and Economic Opportunity," *Essays in Transportation Economics and Policy*, J. Gomez-Ibanez, W. B. Tye, and C. Winston (eds), Washington, DC: Brookings Institution Press, 1999, pp. 437–466.

[15] Suzuki, P. T., "Vernacular Cabs: Jitneys and Gypsies in Five Cities," *Transportation Research*, Vol. 19A, No. 4, 1985, pp. 337–347.

[16] Gerondeau, C., *Transport in Europe*, Boston, MA: Artech House Inc., 1997, pp. 216 and 235.

[17] Gerondeau, C., *Transport in Europe*, Boston, MA: Artech House Inc., 1997, pp. 84, 85, 252.

[18] http://weber.u.washington.edu/~jbs/trans, Website of Innovative Transportation Technologies, Jerry Schneider, University of Washington.

9

A New Kind of Minicity

Variety is the spice of life.

Anonymous

We hypothesize that there may be a way to evolve a high-density minicity that many people might actually prefer to suburban living, an enclave in which the only role of the car is outside its confines. We shall call this new urban environment pedestrian-oriented minicity (POM). We also thought of high-density carless enclave for living and working, but we couldn't pronounce HDCELW.

Our story starts with the shopping mall.

People, at least most people, like the big, fully enclosed shopping malls with their large department stores and small specialty stores, with restaurants and cafes, usually movie theaters, and sometimes a supermarket and a drug store. They seem to like the excitement and the variety that the mall developers and managers do their best to provide. And they appreciate the security. Malls not only draw the shopper and sometimes the diner, but they also draw the folks who want an indoor place to walk or just to people-watch.

We are beginning to see upscale hotels integrated into some malls. These relatively new developments must reflect the judgment that the easy availability of the mall's features, services, and ambiance are a drawing card for the hotel. And surely the mall developers see the hotel as one more source of clientele for the mall. Symbiosis.

How many people would like to live in such a mall? More precisely, how many people would like to live in an apartment or condominium complex that opened into a large, modern shopping mall? It would seem to be a small step from having a built-in hotel to having a built-in apartment building. It would certainly offer convenient shopping for the residents, particularly if the mall included a supermarket and a drug store. A mall would also offer a year-round, climate-controlled environment that is pleasant, reasonably secure, interesting, varied, and even quietly exciting.

At least in the beginning, any such development of, say, a large apartment complex integrated into a mall is likely to be a reasonably upscale development. We suspect that the at least moderately affluent elderly are likely to make up a large proportion of the initial group of people who would be attracted. These are folks who might prefer not to be dependent on their cars for the ordinary chores of daily living and can afford this approach to solving that problem. It might also appeal to many people who no longer feel they need a backyard for their children, who are tired of having to get in the car every time they discover they forgot to pick up the frozen yogurt on their last shopping trip, and those who have learned to hate their lawn mower.

But most people who might like to lessen their dependence on their cars do not want to give them up completely, implying the need for convenient accommodations for these cars. There would also have to be entrances through which deliveries of a new refrigerator or dining room table could be provided. Given the car is still available for access to the outside world, having this inner world also easily accessible should broaden its attractiveness to many people of all ages.

What might be the larger significance to our evolving metropolitan areas of such a marriage of a residential living complex with a mall? Disappointingly, the answer is, "Not much."

The numbers of people involved is just too small. The suburbs would never notice they had left, and they would be greatly outnumbered by the mall's conventional customers. The mall developer might see some small advantage in being able to promise an on-site clientele in signing up shops for the new mall. But he has to trade this off against alternative uses for the space it might take and the investment it would require. An apartment complex developer might, on the other hand, view it as a real drawing card. We suspect that a collaboration is the most likely path to bringing such an arrangement to fruition.

Why this negative judgment? If mall living works for a few people, why not just add more apartments and expand the range of mall services to things like schools, day care centers, and tennis courts to appeal to a broader market?

The problem is space. It is also human behavior. Joel Garreau in *Edge City* gives what he calls the mall developer's first and second laws [1]:

1. An American will not walk more than 600 ft before getting into his or her car.

2. Assuming using a shared driveway, parking an automobile takes 400 square feet.

Two normal city blocks are approximately 600 ft. This number is not carved in stone, and lots of malls go a bit beyond the two-block limit, but without doubt how far people are willing to walk is a real constraint on the size of malls. And as the size of the mall grows, so does the need for parking. Adding apartments doesn't change these rules of thumb and, in fact, creates one more demand on footprint space, with dwelling space competing with space for more shops and perhaps with space for parking. Introducing multistory or underground layers of parking is an option, but it adds a complication and an expense.

So combining living with mall shopping may be a great success for a few, but dynamic growth in this option would appear to be foreclosed unless we can do something to tap into footprint space or existing apartment complexes outside the mall to outgrow the constraint imposed by people's reluctance to walk long distances. We need to permit this minicity that we are trying to create to reach out beyond the confines of the mall itself. Not everybody wants to live in a mall, and even a big mall is no substitute for the full-service minicity we hope to become.

The only possible way we can imagine doing that is to add some form of internal transportation to the mall that is capable of reaching beyond it, something that can form the transportation backbone of a diverse but integrated urban complex of living, working, schooling, worshipping, and playing.

Moving sidewalks, which are common in airports, immediately come to mind. And they may play a role. We visualize something different, however, something vaguely reminiscent of the so-called PRT systems that were studied and tested fairly extensively in the 1970s and are still garnering interest.

These PRT systems were originally envisioned as a network of tracks carrying small, two- to four-passenger "cars" anywhere in the network without stopping. The tracks were typically planned to be elevated, running through the already existing city streets. Since they were outdoors, every car was enclosed and carried its own heating or air conditioning equipment, surveillance cameras to ensure security, and equipment to cope with emergencies. Not surprisingly, such systems turned out to be expensive. There were a few fairly limited systems built, but none that really reflected the original conception of a citywide network [2, 3].

For the POM we would make a few changes. First, the "rapid" in PRT is inappropriate; there is no real reason to go any faster than four or five times walking speeds, perhaps under 15 mph. At 15 mph, a five-min trip carries one over a mile. The need to worry about 600-ft walking distances disappears.

In addition to being slower, we will design our new system to be an indoor system so that there is no need to air condition and heat the individual cars. In fact, there is no need for enclosed "cars" at all. Designers don't have to worry about designing for snow storms or freezing rain or elaborate emergency exits. Out in full view, the security problems are minimized. They are moving at such a low speed that a moving platform with a fence around it for safety might be enough.

We will want capacity for perhaps four passengers plus their shopping bags. There may even be a mix of car types: some might consist of just a few seats, some larger, maybe even some with a few chairs and a picnic table—one's imagination can easily run away. System designers will have considerable flexibility in the choices made. But keeping the passenger load small would permit more frequent service and nearly express delivery for everyone. We envision fairly closely spaced loading and unloading points that are off-line so that a stopped "platform" doesn't impede other "platforms" passing that point.

The technology required for our new pedestrian transport system (PTS) is familiar and readily available. The structure should be lightweight, the electric-powered "cars" would be energy efficient and emission-free, and the controls automatic. The big trick, as we see it, will be primarily the problem of devising a spatial layout that fits with its environment. It will take a certain amount of cleverness to design something that provides the service we envision without taking up an inordinate amount of space. But we have little doubt that it can be done.

The PTS removes the spatial constraints of walking-only, and this changes all the ground rules. First, it lets us expand from a MallWorld to an indoor minicity, our POM. The original mall becomes the seminal nucleus of a much larger living-working-shopping-schooling-worshipping-playing network; with even relatively slow transportation there is almost no real limit on size—we can think in terms of several miles instead of several hundred feet. It also permits malls themselves to be much larger (this may be enough to motivate some mall developers to contribute toward developing our new PTS).

With our minicity we create a large indoor urban enclave that is attractive and dynamic and can accommodate hotels, dwellings, schools, churches, businesses—the full gamut of urban necessities and amenities. The movement within the network is all indoors, but there could be access points to the outdoor world of parks, playgrounds, and living complexes that can now cater to a wide range of income levels. The automobile interface is through distributed indoor and outdoor parking, interfacing with the external road system and the rest of the urban area.

Over time, one can envision new hotels, dwelling complexes, and other kinds of activity centers being added to our growing minicity, designed around the PTS that integrates the whole. As the number of inhabitants grows so do the types of services that can be provided; and as services expand, so does attractiveness to both potential dwellers and outside patrons. The POM may facilitate improved security and living-aids such as the opportunity for more convenient child care for the working mother. The automobile becomes a requirement only for trips outside the reach of the PTS.

It seems to us that one would not have to be elderly to appreciate such a living environment.

These low-speed pedestrian transport platforms might even be fun to ride. Passengers are in the open in full view of all mall activity and other activity centers along the route. They may run directly through the lobbies of hotels and the center courts of office buildings and through parks and campuses in transparent tunnels. (As an aside, we note that such a PTS might have many uses other than as a backbone for a new kind of urban living. Our concentration here, though, is on a hypothesized urban environment that might successfully compete with the alternative that has been the dominant winner through all our lifetimes: the low-density suburb.)

Our indoor minicity would interface with urban transit. As we have said repeatedly, transit is best when large numbers of people are going from the same place to the same place; with lots of riders it's affordable to have a train or bus come along every few minutes and still have a reasonable ridership on each vehicle. But in our low-density metropolitan areas heavy streams of riders are not typical: mostly people start at different places and go to different places. Without the high-frequency service that comes with more concentrated patterns of movement, it's hard to attract any rider who is not forced to use the system. We have reshaped our cities so that the automobile is by far the most economical vehicle for serving its low-density flows. Our new high-density minicity moves against this trend by concentrating passengers.

Now all this may sound very marvelous, but how can it come about? Our crystal ball gets a tad hazy at this point. Will some entrepreneurial company decide there is a market for a PTS and invest in its development? Or will some mall developer decide that a supermall has such great potential that they will fund its development?

Or will our new PTS be funded by the developers of a major new apartment complex on the rationale that if the complex can be directly connected to an already existing mall or office complex or whatever that higher rents can be charged? Perhaps they would also use the new transport system to connect to the common amenities: the tennis courts, pool, and clubhouses, for example. In fact, they may decide they can gain not only from higher rents for providing such conveniences but by selling commercial space in the connecting passageways.

Or, almost as an aside, we mention an amenity that is within our imagination but probably beyond our current technology, at least beyond it at an affordable price. To keep our pedestrian clientele from having to carry their purchases, we speculate an internal "freight" system that can deliver groceries and small packages from "mailing points" scattered throughout the minicity, perhaps directly to dwelling units. The same service could be provided to all access points to the outside, like in the parking structure or the transit stations or adjacent living complexes. The security problem could be handled with codes.

What we are imagining here is much more a conjecture than a prediction. But almost every day brings a new reminder of the ingenuity exhibited within our free market economy in spotting and exploiting new market niches. If the notion we've presented here does, in fact, represent an exploitable market niche, then the odds are good that someone will find a way. And seminal niches—like mall mobility—often have a way of becoming full-fledged markets like a minicity mobility system. There is ample opportunity for many forms of entrepreneurship.

References

[1] Garreau, Joel, *Edge City: Life on the New Frontier*, New York: Anchor Doubleday, 1991, pp. 117–118.

[2] Fabian, L. J., "The Exceptional Services of Driverless Metros," *J. Advanced Transportation*, Vol. 33, No. 1, 1999, pp. 5–9.

[3] Didrikson, P. V., and K. Nickerson, "The Supply Side of the Automated People Mover Market," *J. Advanced Transportation*, Vol. 33, No. 1, 1999, pp. 17–33.

10

The Second-Story City

How to make a silk purse maybe?

A very dependable symptom of a healthy downtown is crowds—lots of people and streets full of cars and buses. The traffic engineer who is trying to keep that traffic flowing has a slightly different perspective: getting rid of all those people would make his job a lot easier.

He's tried lots of schemes. He has set up the lights at the corner so that traffic is stopped in all four directions for a minute or so, and just given the intersection to the crowds to cross anyway they want to. He's tried to let the people go with the vehicular traffic: when east-west cars are flowing east-west people are also flowing. This one requires being very careful about letting cars make turns—traffic flow is important, but it's also nice to minimize the loss of pedestrians. But coordinating lights to serve cars going straight, cars turning right, cars turning left, and agility-challenged pedestrians is no pushover.

Calming the flow of traffic by narrowing the street or by restricting traffic to buses and streetcars are other approaches. Another is to seal off streets in order to create a kind of downtown mall. It was tried for Chicago's famous State Street, but it didn't prove successful. In general, success has been mixed.

Now a few cities, particularly northern cities like Minneapolis and Calgary, have done something that has made our friendly traffic engineer's life much easier: they have supplied enclosed crosswalks that go from the second story of one building across the street to the second floor of the building there—"skyways." Maybe their primary motivation was to let the pedestrians get out of the 0°F weather, but the traffic engineer is grateful just the same.

There can be many blocks of the downtown all tied together at the second story level. The buildings themselves can be rearranged so that the second floors are the pedestrian floors, with walkways through the building, and shops and little interior "minimalls" for their shopping convenience. Once inside this complex, a pedestrian needn't come out until they are ready to leave or they hit the limit of the converted area [1].

The result of this arrangement is a more comfortable environment for the pedestrians than the sidewalk and a lot fewer people trying to share street space with cars and buses.

We described this as an arrangement that takes most of the pedestrians out of the streets and puts them at the second-floor level. We can see no fundamental reason that the pedestrian floor couldn't be the seventh floor instead of the second, or the fourth, or even different floors in different buildings. The latter would complicate things by requiring escalators or elevators, but the potential flexibility is there. Putting everything on the second floor has the advantage of easier access to the street level and standardization of layouts.

The basement level is a possibility. It can work, as examples in Toronto and Chicago and Crystal City show. But we suspect that better light and a feeling of openness and views favor higher floors.

Let's push the envelope

So far we've just laid the groundwork. In the last chapter we described a PTS specifically for interior spaces and in "tunnels." It strikes us that this system would fit right into the second-story city.

It should have several impacts. First, we suspect that the 600-ft walking limit that applies to mall design also has some relevance here, that the scope of any particular pedestrian in terms of city blocks covered is constrained by either that pedestrian's propensity to walk or by the time required to do so. The ability to expand that scope by having available automated transportation should make the whole complex more attractive. And because it attracts more people, a location in that developed area should be much more valuable to any merchant or business that interacts with the public.

Second, because it makes moving around the developed area much easier than walking, it makes the point of entry less critical. Rather than the person hunting around for a parking place near the particular destination they have in mind, they can park nearly anywhere in the general vicinity. Rather than requiring parking structures at closer intervals throughout the developed area, it becomes feasible to think in terms of perhaps a few larger parking structures near the edges—the PTS can provide easy and reasonably quick access to any point inside. This eases the where-to-park decision process for the shopper and reduces the number of cars that have to drive into the downtown region.

Revitalize our downtowns?

Our downtowns represent a very large sunk investment. They are valuable because the proximity they afford their inhabitants is apparently important to many kinds of commerce. But they have had a tough time competing with the suburban mall, with its park-once and have access to all kinds of stores, movies, restaurants, and other services.

We said sunk investment, and a rational fellow, such as our Martian friend, might say, "*Just because it is there doesn't mean that it has to be used. It is just like that old, energy-inefficient furnace you used to have. You did the right thing when throwing it away. The downtowns are similar. If they can't cut the mustard and swim in our competitive world, let them sink.*" (The Martian himself sometimes sinks under the weight of mixed metaphors.)

That is a good point, but we are thinking of trying to make them more attractive so that maybe they can compete better. Moving in the direction we just suggested would seem to integrate the many diverse buildings that make up a large downtown into the first cousin of a shopping mall, with heavy commercial overtones. In fact, the diversity of activities, architectural styles, and services could very well exceed those offered by the typical mall and, therefore, make it even more attractive.

Being downtown, parking is likely to be in an enclosed, multistory parking structure rather than the typical large-area, outdoor parking lot of a suburban mall. This would certainly be a more attractive option when weather is bad. The availability of a PTS connecting to the parking structure provides easy access to any point in the whole developed area.

It would certainly seem to be worth some serious thought.

Reference

[1]   Robertson, Kent A., "Pedestrian Strategies for Downtown Planners: Skywalks Versus Pedestrian Malls," *J. American Planning Assoc.*, Vol. 59, 1993, pp. 361–370.

11

The Variegated City

Variety is the spice of life.

Anonymous

The minicity and the second-story city are examples of some of the various paths along which cities might evolve—just examples, one of a kind. The automated transportation that might aid in remaking downtowns is the same sort of thing—just one of many possibilities.

Indeed, we can make lists of many ways to go.

- Using the neighborhood car as the transportation backbone, configure neighborhoods sequestered from through traffic.

- Because cars and parking garages can be automated, store lots of cars off the streets and free up some street spaces for neighborhood gardens.

- Delivery vehicles can be automated too, so let's have most all deliveries made at night, placing merchandise in lock boxes by our door.

- Improve our ability to build short simple tunnels and improve access to off-street parking and go under historic districts.

Making lists is easy, and it is easy to tell others to make lists. But it is sort of like reaching to the shelf and dumping ingredients into the pot willy nilly. Does cooking have some rules, a recipe? Any recipe must aim to please, so what are we trying to cook?

This chapter will take up this question. We will look around at the way things are going, looking for things that fit. In the chapter to follow, we will consider roles of innovations big and small. Using a two-step dance metaphor, we will indicate how innovations translate into enriched options for living. Finally, we will mention the context for innovation and other factors to be considered in the search for innovations.

Cities thrive by variegating

Cities thrive by satisfying their inhabitants. And inhabitants come in all flavors, different interests, with different kinds of jobs in different economic strata, different cultures and religions, and different preferences in living styles. The city itself represents a convergence of processes [1]. These differences are not accommodated by homogenous cities.

The walking city of times past could not satisfy the modern suburbanite; it took additional forms of transportation to provide the suburban option. Transportation and communication technologies have weakened the "close by" proximity requirement that constrained these alternative options in the older cities; now, free of that requirement, cities increasingly have opportunities to accommodate a much greater range of preferences for working and living.

A succinct way to put it is to say that cities have become increasingly variegated over the decades as new forms of transportation and communication have enabled cities to accommodate the varied preferences and requirements of a diverse population. Folk want their own thing—recreation, work, schooling, housing, and friends—and specialization in all things has been an overwhelming social trend [2].

Not all eyes see things the same way. There is much railing against the "sameness" of suburbia and the advantages of the tight knit city where one can encounter diversity and have their wants satisfied by just walking around [3]. We don't argue with this view. It is an environmental design perspective, one of the many forces that will help shape the city in its continuing evolution. But we have a larger point in mind.

Let us try to momentarily view the situation from afar. We see some self-contained enclaves or urban villages that some folk like; we also see some that are distinctly less attractive. In some places we see what looks like blandness. Yet we see transportation and communication widening peoples' worlds by providing both better information and better physical access to alternative ways of living, allowing and motivating folks to transcend the boundaries of suburban, racial, in town, or rural ghettos, enabling all to grasp opportunities for variegated living. That is how transportation helps shape cities; it enables the realization of desires for change.

What do we want of our cities? We want them to be both permissive and supportive in accommodating their inhabitants. This is a dynamic process: life is not static. Peoples' tastes and imperatives and means continue to change, and the successful city will be dynamically variegated in response to those changes—or even, through the ingenuity and foresight of the citizenry, able to preempt and precipitate these changes, offering possibilities for completely new modes of living. The successful city is a work-in-process without end.

Looking ahead at trends—folk living everywhere

Transportation innovations ought to be consistent with social and economic trends. For one thing, transportation and communication improvements are changing the meaning of distance and the isolation of places. How do these changes interrelate with social trends?

Years ago a few large cities had near monopoly on the easy transfer of information, diverse jobs, education, varieties of recreational and social-cultural activities, and the other things we take for granted in modern urban society. Indeed, we still think of New York, Paris, and a handful of other places as "having it all."

But modern communication and transportation technologies, along with education and many other things, have eased that monopoly. No matter where one lives the advantages of cities everywhere are increasingly at hand. Actually, society has been chiseling away at the big-city monopoly for a long time. For instance, widespread political support for the introduction of air mail was based partly on the belief that it would reduce the communications-information advantages held by New York. Today the wiring of schools to the Internet picks up the theme.

Cities still get advantages from size or scale. For instance, a fair-sized city is needed to support a zoo or a symphony orchestra. But some of the old advantages of size are democratizing down the size scale. The stockbrokers and theaters, for example, in Des Moines compare much better today than yesterday to, say, those in Chicago.

One way in which cities become more efficient is through specialization; size also impacts the ability to play this role, as does the heavy hand of history (historic path dependence—by sheer chance or for some other reason, something got started there, and it persists). Among other things, Nashville has country music and firms managing health care facilities, and San Francisco and Boston have high technology. Agglomeration advantages occur in these cities as similar activities share information, supporting services, and other production factors. History, size, and agglomeration economies are all at work.

Today's information and transportation advances are changing the meaning of place and loosening the bonds of scale and historic development paths on location. Improved communication and travel opportunities are permitting a few people, no matter their location, to begin to live in virtual cities.

How we live—is everything old new again?

The variety of services continues to expand, a trend that we think is all for the good. Here we describe a few specifics.

After the Civil War the railroad opened up the farmlands of the west, expanding from some 35,000 miles of track in place in 1865 to nearly 200,000 in 1890. Much of this was built under the assumption that "if we build it, they will come." And they did. But in a wonderful example of the

Law of Unintended Consequences, the Homestead Act of 1862 almost guaranteed that the result would not be a hinterland dotted with rural villages and small towns, with their stores and churches and schoolhouses, but a farming hinterland dotted with individual farms, separated often by miles from its nearest neighbor. The Act specified that a settler had to live on his land for five years to assure his claim, so the settler was constrained from just moving into the nearest village and going out every day to farm his land; he had to put his home on the land. Indirectly, the Act invented Montgomery Ward and Sears and Roebuck [4].

These rural folks, isolated from each other and from easy-to-reach stores, almost had to depend on catalog shopping, using the mail and parcel post system and rural free delivery (RFD, established in 1893). This became the heyday of catalog shopping, a period that predated the widespread use of the automobile and truck. Pouring through a catalog, the shopper could choose from farm implements, clothing, and many other things, including for a time prefabricated houses. Many even tried to shop for wives this way, with more mixed success we are told.

By the 1920s the picture was changing. Now about half of the population lived in urban areas, transit was already serving the customers needed to support downtown shopping, and outlying shopping centers began to develop at places on transit lines. In rural America, the automobile and the truck began to make trips to shops at the county seat practicable. As the number and quality of automobiles increased and roads were further improved, impacts continued to run their courses, and catalog shopping lost much of its market. A myriad of shopping opportunities emerged—auto rows, large shopping centers, megastores. A look at a telephone directory confirms the large number of goods and services available.

Today, catalog shopping has made quite a comeback, as a convenience for the urban dweller rather than as it was 100 or more years ago, a necessity for the rural homesteader. We suspect that lots of people don't find the driving, the parking, the crowds, and the shopping as much fun as others seem to and are happy to avoid the hassle. And catalog firms, offering as they do a very wide range of products specialized to different tastes, offer more choices to the shopper than is easily available in any city, and lacking completely in small- to medium-sized ones.

It is convenient. Ordering can be done by phone or computer over the Internet. The use of credit cards, other credit, and check writing takes care of the monetary requirements for transactions. If purchasing out of state, sales taxes may be avoided (states will, in time, probably close this door). Order, then just wait for the goods to be brought to your home. And most firms make it easy to return items that didn't suit, taking much of the sting out of the big disadvantage of catalog shopping: the inability to see the item in the flesh before making a purchase.

Catalog firms have responded by concentrating on collections of products specialized to market segments. Communications and information technologies identify and track purchases by item and usually direct delivery from factory to home, almost eliminating warehousing by the catalog operation. Their "business model" is very different from that of the conventional retailer, a difference that has been made easy by the computer, the fax, and the Internet. Transportation—specialized to the role—is provided by a variety of parcel delivery services.

There is a parallel story in the procurement of services using the "Yellow Pages," industry directories, and other information bases. Services range from temporary employees, to tutors for the homes of children with learning problems, to rain gutter fabrication and installation, to almost anything one can imagine. In addition to communications, these service activities require adequate transportation services and, in many cases, specialized transportation services. Equipment or materials may need to be transported. The rug-cleaning people bring their equipment on a truck, and the automobile detailer brings cleaning equipment and supplies to the home or wherever the car is parked. One can get a broken windshield replaced from a van.

We already conjectured about the implications of automation of many of these functions. In particular, we described little delivery bugs that would be cheap to own and operate, but that is far from the only models.

Chauffeuring

The wizardry of the computer could bring new dimensions to the chauffeuring function. Today, there is walking to school, the school bus, parents sharing driving, older children driving, and other arrangements. Tomorrow an organized chauffeur service might be developed using computer-based scheduling. We can see the Amalgamated Kiddy-Ride Service, bonded and regulated to ensure safety, or a "grandmother-ride" service, or other variations of chauffeur-like services. Many of these variegated services might just evolve from increasing specialization of taxi services—now the jack-of-all-trades.

Full automation makes it even easier. The family car picks up John at the school and takes him to his dental appointment, and on that same trip the car cycles by other schools to take Jane and Sally to their karate lessons. The automated car goes home or somewhere else until time to pick them up. No hands!

We tend to get a bit carried away. We are only trying to make the point that we have only seen the beginning of specialized services, tailored to nearly every taste and preference. The technologies are available, their familiarity is spreading, and their versatility is increasing and coming down in price. Mix well with entrepreneurship, and urban life takes on a richness in material things that are well beyond anything ever available to any king in history.

How we work

In all of this our focus has been largely on the functions of living, not on the functions of making a living. We talk loftily about accommodating preferences and tastes and seem to gloss over paying the bill. The fact is that a fundamental function of the city is commerce in all its facets. Professor Howard Saalman, in his book *Medieval Cities*, put it somewhat more bluntly than some might like, but in our judgment he put it accurately [5]:

> "A city is a tool for the production and exchange of goods and services. A city may also be a place where people live, study, play, worship, or have children. It may be a place of magic or terror, of beauty or of ugliness. But such things are true of other places as well: fields, mountain tops, or caves—such attributes are subjective and secondary to the essential function of a city."

This characterization, made perhaps while thinking of a crueler, fight-just-to-stay-alive era, still carries truth for today. After all, the economy supplies the wherewithal to support all the rest. Economic health is a necessary condition for a successful city, although not always a sufficient one. By providing proximity, cities enable more efficient production, distribution, and consumption. Over the years, folk have migrated to cities to take advantage of the jobs created by its efficiencies. We want to always have efficiency in mind, and much should be made of innovations that improve general productivity.

If we look back to the 1920s when urbanization was at its most rapid pace, what we think of as "doing better" was extolled in the popular literature. Novelists were describing how young people of the day were finding their destinies in urban environments, finding better lives. But first and foremost, migrants went to cities to find work—the better life came later.

Dickens was critical of the "dark, satanic mills," the really awful working and living conditions in London in the early years of the Industrial Revolution. What he did not consider was the "compared to what?" People kept flocking to the city to work in those same dark, satanic mills because they still offered the best option available to them. Life then was very hard, but the city was, in fact, carrying out its primary function of an economic tool.

We have moved far from dark, satanic mills. We are a far richer people than lived in 19th-century London—or anywhere else, for that matter. We can afford to pay more attention to the ambiance of the city, to invest more in making it a pleasant place for people, in Professor Saalman's words, to "live, study, play, worship, or have children." In fact, as cities compete for skilled labor, all those characteristics we bundle in the word ambiance are an economic asset.

We Americans have also become an urban society. Today, mostly everyone lives in an urban context. Population statistics tell us that about 90 percent of us live in towns, maybe small towns or middling cities, many in suburbs, and still some in old central cities and in downtowns. (About half of the 20% of the population classified as rural by the census lives in urban settings.) And we live in urban styles, as we remarked earlier when discussing the breaking of big city monopolies on information and varied job and social opportunities.

Not too many years ago the Congressional Office of Technology Assessment undertook a far-reaching analysis of what it called the

American transition [6]. It looked at high- and low-technology industry, farming, education, and other activity sectors. It used a language of production and consumption recipes, and running through recipes were themes captured in the expressions: more flexible, knowledge-based, service sector growth, communications and education, and specialization.

Matched with those themes were comments on skilled work forces and continuing education, increased participation of women in the work force, science and technology as an input to production, and the changing nature of job tenure.

Reading reports such as this one and looking back and around, more and more of the old image falls away, the image of centralized production in a large facility achieving economies of scale. It falls away in favor of shifting markets and products, expanding and contracting work forces, and shifting skill mixes.

While many old arrangements continue to hold, looking ahead one increasingly sees a highly skilled work force moving here and there as demand for skills shift. That is required by and matched by fast-on-their-feet production systems that make intensive use of information and communications and that combine labor and capital inputs in flexible ways.

The variegated city

The trend is almost a flood. We are seeing more variety and greater degrees of specialization creeping into how almost every function of our lives is carried out. More and more, the differences in tastes and temperament of our populace can be indulged because there is a wider variety of options and opportunities available to indulge them.

Now we admit that justification for this observation is largely anecdotal; we know of no careful accounting to quantify this trend.

Our basic point is that we are a variegated people, with differing talents, preferences, interests, needs, and energies. And we are finding that our cities are evolving to provide an increasing menu of options to cater to these differences. A variegated populace leads to variegated cities, variegated in services, in infrastructure, in ambiance. Our parochial view forces us to observe that this variety is aided and abetted by flexible and diverse capabilities in transportation, a subject we turn to next.

In the chapter to follow we will consider roles of innovations big and small. We will indicate how innovations translate into enriched options for living. Finally, we will mention the context for innovation and other factors to be considered in the search for innovations.

References

[1] Berry, Brian J. L., *Human Consequences of Urbanization*, New York: St. Martins Press, 1973.

[2] Moore, W. F., *Social Change*, New York: Prentice Hall, 1963.

[3] Katz, Peter, *The New Urbanism: Toward an Architecture of Community*, New York: McGraw-Hill, 1994.

[4] Boorstin, Daniel J., *The Americans: The Democratic Experience*, New York: Vintage Books, 1974.

[5] Saalman, Howard, *Medieval Cities*, New York: George Brazillar, 1968.

[6] Office of Technology Assessment, *Technology and the American Economic Transition: Choices for the Future*, Congress of the United States, 1988.

12

Variegated Transportation for Variegated Cities

The problem most often addressed is how to administer the status quo. The more interesting and fruitful problem is how to improve it.

Joseph Schumpeter, *Capitalism, Socialism, and Democracy*

When we described the evolution of the city, it was easy to spot the really blockbuster transportation innovations. After millennia of nothing but walking and a few muscle-powered vehicles, the electrification of the city and the electric streetcar were major changes. Then came the really big one, the internal combustion engine that spawned the automobile, the truck, and the bus. Many people would add the diesel locomotive and the jet engine; these are more intercity than urban, but it's a definitional issue with which we wouldn't argue.

135

Should we recognize things like urban public works departments and the development of traffic ordinances as blockbuster innovations?

Certainly the fully automated vehicle of Chapter 2 will qualify, and maybe the POTS described in Chapter 9 will some day make the blockbuster list.

But the richness and variety of transportation services we enjoy today stems not only from the blockbusters, but from their offspring. Each blockbuster innovation has continued to evolve, usually in very small steps, to produce a host of variants. Some of these result from continuing technical innovations, like the modern electrified rail transit systems that are derivatives of the train and the trolley. But for many—maybe most—of the variant systems the technical changes are trivial, and the additional forms of service follow from the creation or recognition of new market niches: essentially the same hardware being used in a new way.

For innovations big or small, it is the market that is critical. This point cannot be overemphasized. As Aaron Gellman of Northwestern University likes to put it, "An innovation without a market is the sound of one hand clapping." Tying innovations to markets and economic growth, Gerhard Mensch has coined the word "pseudoinnovation" for things of little consequence but much hype [1].

A market niche is a situation in which some reasonable number of people have an unfilled need, perhaps without even realizing it—like TV before anyone ever saw one. History and ordinary experiences tell us that a lot of developments or innovations take place in market niches; in fact, the existence of a latent or existing market niche—an unfulfilled need—is a necessary condition for a successful innovation.

So the wide variety of transportation services available to serve the diversity of peoples' tastes and needs doesn't all stem from blockbuster changes but also from incremental innovations in either the existing hardware or in the way it is used or both in response to or in anticipation of some market niche.

The fact is that even our "blockbusters" are only blockbusters in hindsight. The first Model-T Ford emerged only after literally hundreds of prior auto designs had been built; and the first commercially successful airplane, the Douglas Corporation's Model 3 (the DC-3), appeared roughly 30 years after the Wright Brothers, first flight at Kitty Hawk.

Earlier models of both of these marvelous transportation devices were technically inadequate to create the market niches they were destined to fill. In both cases it took a very long succession of incremental improvements—innovations—to cumulate into a technically and economically successful model.

The point of this little discussion is to remind us that future improvements and diversification of transportation services does not just depend on finding new blockbusters. The progressive improvements in transportation yielding variegated transportation as a crucial element of the variegated city are not single events but the product of a cumulative series of innovations, very frequently small and incremental innovations.

The electric streetcar is a case study in the innovative process.

Electric streetcars and the two-step dance

It is conventional to say that innovations occur when someone recognizes a market niche and acts to fill it. Often it works the other way: a new variant, an innovation, is introduced to serve one niche, and its availability triggers new needs and creates new niches and ultimately new services. We made this point before when we stressed the enabling aspects of transportation; here we describe it slightly differently.

It is helpful to think of transportation-induced change as a two-step dance. New or improved transportation services let people continue doing what they were doing a little cheaper or a little faster or a little more comfortably. That's step one—cheaper, faster, better services. Often the batting average is two out of three, say, faster and better but not cheaper.

Step two is when someone says that now that we have this new transportation, we can start doing these old things differently or do entirely new things. This often happens to the transportation itself. People find ways to use it that go well beyond the use for which it was originally intended, and a new transportation service is born (Figure 12.1).

The example of electric streetcars illustrates this "dance." The seeds that enabled the shift from horse- or cable-drawn cars were the earlier development of electrical power generation and distribution and electric motors as well as experience creating political and institutional

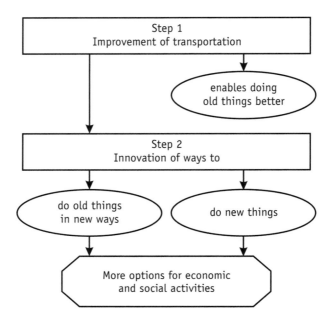

Figure 12.1 Two steps to increasing social and economic options.

arrangements for franchising, providing, and financing services. Cars were increased in size, decreasing cost per seat mile, and service was faster and more reliable. That was the "improve transportation" step, and it scored well on cheaper, faster, better scales. The market clamored for services. Even in Los Angeles where today the automobile is king, "Big Red" streetcars found a ready market [2].

But the big payoffs were from the ways that improved services combined with other things to improve urban life. This is an important point that, although obvious once said, bears underscoring.

The new services found a ready market in growing crowded cities where improved access to jobs, recreation, shopping, and new residential areas were much needed.

Suburban living was far more than the electric streetcar. It was the result of the confluence of a lot of things: the availability of land for development; the ways in which land could be subdivided and sold; new processes for housing construction and new materials; the financing of home purchases; and new and old protocols for providing urban water supply, electrical and communication services, and possibly others that we have overlooked [3].

Combining the elevator with ways to construct tall buildings—and light and heat and cool them—gave us high-rise downtowns for working, shopping, and recreation. The streetcar connected those downtowns with workers and customers traveling longer distances than before.

Looking back, the introduction of the electric streetcar and the response of the city to this new service seems seamless and simple. And in many ways it was because the conditions were ripe for its introduction.

Responding to the public's clamor for services, city governments willingly franchised private companies to seek capital, built properties, and operated profitable services. Governments already knew how to franchise and regulate services because of previous experiences with railroads. The existing street infrastructure provided spaces for laying rails and providing services.

Narrowing our view to particular cities, streetcar services emerged in market niches. Franchises resulted in a line here and another there as entrepreneurs could imagine services, raise capital, overcome the resistance of horse-drawn tram operators, and make deals with governments. Promoters sought profits, so streets and routes with high-potential ridership were first to have service.

As new things to do were found, service providers responded, and the streetcar system evolved to accommodate. Special funeral cars and services were created, for example. To serve Saturday afternoon and Sunday travel, schedules and equipment were configured for recreational and leisure travel. Cities developed streetcar-accessible parks, and some recreation facilities were provided by the streetcar companies.

Early on, Coney Island was a small, ferry-served resort on the shore of Long Island. Inexpensive streetcar service greatly increased the number of persons who could visit and developers responded by creating amusement parks, each park somewhat specialized to market segments—one specializing in thrill rides and another in the arts, for example. Stories similar to New York City's Coney Island story were repeated in many places [4].

Hindsight tells us that the streetcar worked technically and economically as a cheaper, faster, better substitute for horse-drawn trams and long-distance walking. It also tells us that streetcar services enabled arrays of social and economic innovations that improved the efficiency and quality of urban living.

There were bumps in the road. The entrepreneurs did not always properly gauge the various niches they were trying to fill. Indeed, the first application failed, and *The Railway Age* reported in 1889, "After being tried for 18 months a street railroad in Richmond, VA, which has been operated by electricity has been declared a failure, and the company will go back to horse or mule power."

Even in the heyday of streetcar services circa 1890–1920, many companies were not profitable. By the 1930s many of the enterprises energized by streetcar services were also in fiscal difficulty. Real life is a harsh taskmaster; and when enthusiasms fade, government support may fade in favor of other modes of travel. Paul Barrett tells us how support for streetcars faded in Chicago when the automobile became popular [5].

Public–private roles in how things happen

It is often said that the private sector is full of entrepreneurs looking for market niches for new products or services. While the "full of" may be an overstatement, there are certainly lots of folk looking for market niches that portend a profit. Entrepreneurs are motivated to find new and better ways for people to live—all kinds of people. For them the acid test is, Will people like the service or product enough to pay for it? Private sector entrepreneurs have no choice: they must march to the drum of the market.

And there is the public sector, and it is big. In much of the developed world, public sector expenditures are about equal to private sector expenditures. The ratio is less for the United States. But the simple citing of fractions ignores the complex interrelations between public and private. Looking for public–private interrelations with transportation innovation in mind, it pops out that governments provide the infrastructure-context for innovation. Let's dig into that.

Infrastructure sounds like a technocrat's word, and as it is used it brings up images of physical things—school buildings, sewage systems, streets and traffic lights, and airports, for example. We may properly extend that image to recognize the presence of public sector innovators, although there are some rough spots. Importantly, the tooth and claw of market competition and the make-a-profit test aren't present, so the payoff for the innovator found in private sector markets also isn't present in the public sector.

A broader interpretation of infrastructure extends to the soft sides of things—the ways markets work, definitions of who owns what, rules for trading property and the adjudication of disputes, and ways to reward creative actions. Regulations that say what can and cannot be done are an important consideration. For short, think of the legal, government, and economic infrastructure.

When we consider innovations that might improve urban transportation the soft and the physical urban infrastructures cannot be ignored because they may work in supportive or conflicting ways.

City (and other governments) bring about (or constrain) change using laws of general applicability, such as those underlying tax codes. Governments create agencies with legal and economic powers to pass judgment on existing or proposed activities, for example environmental or public utility regulation. Some are planning agencies and others make investments and operate programs such as urban street and traffic management programs.

Planning agencies play two kinds of roles. The first is creative planning, looking ahead and trying to discern the best of broad paths of future change. The other is basically as an enforcer of rules and regulations. In his book *The Changing Face of Cities* the British geographer J. W. R. Whitehead noted that planning was "a procedure for resisting departure from precedent" [6].

This primitive civics presentation of the obvious would hardly pass muster in a kindergarten. But no matter. The point is that the actions of entrepreneurs must fit or finesse the patterns of constraint and encouragement represented by hard and soft infrastructures. Again, the entrepreneur strives to guess correctly that a service offered within that context will in fact generate sufficient revenue to cover costs.

In this context, there are two aspects to providing an environment for private sector action. The first is providing what the private sector needs in order to act; we will give an example shortly. Failure to do so is a "something missing" dysfunction. The second aspect is the "something is in the way" dysfunction. It has a "we have seen the enemy, and it is us" flavor. We refer to government policies that stymie creating or picking up options.

Excessive economic regulation receives much attention as a something-in-the-way dysfunction. Let's use taxi service in New York City for illustration, a story we have read from time to time in the city's newspapers. Similar stories can be read in most any big city newspaper.

It seems that at the beginning of the Great Depression of the 1930s there was concern that too many taxicabs were chasing riders in the New York market. It was thought for a variety of reasons that wouldn't do. So to control supply, the number of cabs was set at about 12,000 and cabs were to have a medallion (license) displayed on the hood of the vehicle. Many cities implemented similar schemes at about the same time. They also sought to prevent competition to transit agencies by drivers of cars who would solicit patrons from those waiting at bus stops.

Many point out that regulation acquires vested interests and becomes difficult to dislodge. Inefficiency and decreases in social welfare are also noted. A shortfall in the supply of taxi services is indicated by the price of about $250,000 for the medallion for a New York taxi, and those that have medallions resist the city's increasing the number available. It is not surprising that the number of medallions available are about the same as the number in 1930. It is also not surprising that fares are high compared to other cities.

The preservation of inefficiency by vested interests is a legitimate concern. Beyond that, we point out that preservation thwarts entry by new suppliers who might innovate new kinds of services. By resisting change, regulation is the deadly enemy of innovation.

A New Yorker might say that we overestimate the power of regulation. He would point out that there are lots of taxi-like operators who do not have New York taxi medallions. The shortage of cabs makes many market niches available, and operators fill them in innovative ways. So perhaps instead of saying deadly enemy we should refer to regulation as distorting, sapping resources, delaying, and otherwise affecting innovation.

Turning to supportive relationships between the public and private sectors, nowadays there is much interest in public–private partnerships as a technique to lessen private-sector financial risk but still exploit at least some of their market orientation, technical expertise, and risk-taking willingness. This is by no means a new idea. But it has new aspects as we hear of new ideas from time to time as well as new experiences in many nations [7].

There have been public–private partnership actions at the city, state, and the federal level; usually as government sought to promote transportation services in a mercantilism style. The promotions of railroad services by cities, the states, and, later, by the federal government are exemplary of mercantilism [8], as is the case today for air and water ports, highways, and river improvements.

The public–private partnership idea remains popular. But increasingly attention goes to the alternative of providing a technique, a mechanism that lets the private sector act without specific ties to the government. A transportation case is covered in a well-received Brookings Institution book published in 1997 [9].

In that book, Daniel Klein and his co-authors discuss what they call curb rights. The idea is that the soft infrastructure of government—the legal, government, and economic infrastructure—lacks the property rights and market mechanisms needed for competitive services in mass transit. As things work now, government-operated transit systems have rights to passengers waiting at curbs (bus stops). Why not give or rent specific curb rights to other types of operators? In addition to regular streets, curb rights at airports, at baseball stadiums, at parks and schools? There are lots of different ways the infrastructure for the provision of services might support varieties of competitive services. Klein and his co-authors imagine that if rights were appropriately demarcated and traded, conventional transit and transit-like services could be supplied by the private sector and perhaps in more innovative and effective ways.

But it's hard to totally disassociate the public sector from transportation issues, partly because transportation is always out there in plain sight and partly because the transportation experience taught governments much of what they know about regulation, subsidy, consumer protection, project and program evaluation, and lots of other things that governments do. The progressive government movement was boosted by the task of providing transportation infrastructure. A professionalized civil service was one result. Roberts' Rules of Order, widely used to bring order to deliberative meetings, were developed to aid debating waterway investments.

On the downside, transportation construction and operations have provided ways for governments to reward and punish those who are governed—the taking of taxes from here and spending them there, for example. The term "pork barrel" was coined as waterway investments were directed to favored claimants, and all have heard it applied to today's decisions on transportation and other investments.

It is fair to say that the transportation experience has affected the effectiveness of government as well as the balance between those who govern and those who are governed. Already, privacy concerns have been raised about today's intelligent transportation systems. When we think of the many ways transportation and communications intertwine, this is only a starter question. Who goes where, with whom, and consumes what at what price are questions that airlines can already answer.

Tomorrow? We cannot say how things might go. Perhaps seesaw describes the situation. Sometimes more government, sometimes less; more here, less there.

Constrained exploring

We have talked about blockbuster and less sweeping innovations, emphasized markets and entrepreneurs, and noted the roles of public sector. Referring to the two-step dance, we said that an innovation may find its first niche by being cheaper and faster or, better yet, that the big payoffs from innovation come when they permit doing old things in new ways or doing new things. A lot has been said, but two important points haven't been stressed enough.

The word "exploring" sums up one point. We will achieve tomorrow's variegated transportation in the same way we achieved today's, and that is by exploring options and finding markets. The future has to be found by exploring.

Once said, that's sort of obvious. Also once said it explains why our exploring personal transportation, freight delivery bugs, and other things had a "perhaps this or perhaps that" tone.

Exploring needs to be guided or constrained. That is the second point. Social trends, ways that technologies might evolve, available market niches, and limitations of time and money keep us from trying everything—constraints call for insight, reality checks, and such. We are constrained by social and economic realities.

And we are also constrained by energy and ecological realities, topics we will address in the next chapter.

References

[1] Mensch, Gerhard O., *Stalemate in Technology: Innovations Overcome the Depression*, Cambridge, MA: Ballinger Publishing Company, 1979.

[2] Crump, Spencer, *Ride the Big Red Cars: How Trolleys Helped Build Southern California*, Corona del Mar, CA: Trans-Anglo Books, 1975.

[3] Warner, Sam Bass, *Streetcar Suburbs: The Process of Growth in Boston, 1870–1900*, Cambridge, MA: Harvard University Press, 1962.

[4] Adams, Judith A., *The Amusement Park Industry: A History of Technology and Thrills*, Boston, MA: Twayne Publishers, 1991.

[5] Barrett, Paul, *The Automobile and Urban Transit: The Formation of Public Policy in Chicago, 1900–1930*, Philadelphia, PA: Temple University Press, 1983.

[6] Whitehead, J. W. R., *The Changing Face of Cities*, Oxford and New York: The Institute of British Geographers, Basel Blackwell, Ltd.,1987, p 144.

[7] Wright, Charles L., *Fast Wheels, Slow Traffic: Urban Transportation Choices*, Philadelphia, PA: Temple University Press, 1992.

[8] Dobbin, Frank, *Forging Industrial Policy: The United States, Britain, and France in the Railway Age*, New York: Cambridge University Press, 1994.

[9] Klein, Daniel B., A. T. Moore, and B. Reja, *Curb Rights: A Foundation for Free Enterprise in Urban Transit*, Washington, DC: Brookings Institution Press, 1997.

Part III

Energy and Emissions

13

Energy Use and Pollutant Emissions

Our elephant has many parts. Here we look at another one.

Most of the energy used by transportation is produced from carbon-based fuel (overwhelmingly petroleum), and potentially earth-warming carbon dioxide and other emissions are associated with its combustion.

What are the magnitudes of energy use and pollutant emissions? The U.S. Department of Energy publishes the *Transportation Energy Data Book*. This and publications of the U.S. Department of Transportation tell us about government programs, regulations, and trends in fuel consumption and emissions [1, 2]. (Unless otherwise indicated, these references are the sources of data used in this chapter.)

The *Energy Data Book* is comprised of about 200 pages of tables and figures, with interpretive text sprinkled here and there. It presents such data as are available in a "no ax to grind" style. Produced annually, 18 editions have been issued, each fuller than the previous edition. Drawing from many sources and providing text in the languages used by energy experts, the report is symbolic of the growth of what might be called the transportation energy establishment.

The subject deserves attention: transportation uses a lot of energy, about 70% of U.S. petroleum consumption, and about one-quarter of energy consumption from all sources.

Over half of the petroleum is purchased abroad. In the abstract, offshore purchases are of no concern because the workings of markets and exchange rates settle accounts—nations export some things and import others depending on comparative advantages. In practice, however, inflated cartel prices, potential supply disruptions, and the effects of ties to off-shore petroleum sources create military and political concerns that are very bothersome. These considerations describe one aspect of the petroleum consumption problem.

The emissions that are tied to energy consumption are another aspect of the problem. The combustion process produces oxides of nitrogen, carbon monoxide, volatile organic compounds, and particulate matter. These emissions, and the smog created as sunlight works on some of these compounds, have adverse health effects. Sulfur in fuel (now largely eliminated) adds acid to the atmosphere and the potential for acid rain. Lead was a problem until lead was eliminated from fuels. There is debate about other pollutants that should be recognized as hazardous, but most policy attention has been given to the "criteria" pollutants, those just mentioned.

Transportation vehicles also produce carbon dioxide, but because there are also many other sources, its contribution is only about one-third of the total produced by man's activities. Carbon dioxide is a greenhouse gas. That is, carbon dioxide traps heat from the sun by slowing the radiation from the Earth back into space. Therefore, it may be a source of global warming; and global warming has the potential for increasing sea levels, shifting patterns of dry and humid areas, and affecting crop production, for example.

The global warming idea is not a new one—the carbon dioxide heat blanket effect was recognized about 100 years ago. However, the analysis of the phenomenon remains complex, having to consider such things as the dynamics of the carbon dioxide generation-absorption cycle and the balance between the heat-trapping effect of greenhouse gases and the cooling effect of cloud cover, which reflects the sun's heat away from the Earth. The data say that the amount of carbon dioxide in the atmosphere is increasing, but the existence of carbon dioxide—caused warming and its pace is open to considerable debate. And presuming that the Earth is still emerging from the last ice age, there is nothing that says where we are now is the long-run temperature equilibrium. Today's situation may not even be optimum for mankind as a whole.

Because of unknowns, "caution and prudence" is in order, just as George Marsh said in 1863 when considering the impact of man on nature [3]. Marsh is forgotten, but there has been systematic consideration of Marsh's subject. Even more thinking might be useful today when potential problems are of real concern to many policy makers, and an agency of the United Nations has pressed for international agreements putting a cap on and eventually reducing emissions.

How should we view today's situation? Transportation does use energy prodigally, and burning petroleum does produces emissions that are not benign. Some who carefully examine the situation conclude that problems are manageable [4]. There are others who view the situation as already apocalyptic [5], they say: if we don't get rid of cars, if we continue to risk major oil spills, and if we keep putting more and more carbon dioxide in the air, for example. They worry about the development needs of the have-not nations, population growth, wars, stagnation of economic growth, and famine. Some see major conflicts over dwindling oil supplies. If, if, if.

We do not fluff over concerns and fears: the future is unknowable, and no one can guarantee the dire won't happen. We need to be concerned, but we don't want to lose our perspective or do counterproductive things.

In his book *The Coal Question*, written in 1885, W. S. Jevons analyzed the increasing rate of coal consumption in England [6]. Forecasting the exhaustion of coal supplies, he posed the public policy question as business as usual or reductions in production. He said "... we have to make a choice between brief greatness and longer continued mediocrity."

In our judgment, Mr. Jevons badly underestimated the ability of free societies not only to cope but to improve. We do not need to accept sacrifice and mediocrity as a future for transportation or for the world; new technologies and new ways of thinking about old technologies continue to offer new options, even when there are clouds in the sky ahead. Here we will describe the current state of affairs as we understand it and explore some of the possibilities for the future.

Our abundant energy use today stems from the discovery several centuries back that energy could be harnessed to substitute for muscle. Its growing use is intertwined with the not-independent trends of a larger and more prosperous population in the world. These are serious headwinds for significant curtailment of use. Few would vote that we go back to muscle. Even fewer will volunteer to personally contribute to decreasing the population. And most of us favor increased prosperity. We can improve the efficiency with which energy is used but seldom with zero cost to achieve it. Every move we make to lessen energy consumption or reduce its undesirable side effects involves a tradeoff.

We should recognize that over the last 20 to 30 years there has been great progress in the United States: the air is cleaner in many regions, and the rate of increase in energy consumption has been slowed sharply. But the future portends increasingly motorized transportation in industrializing nations, and the estimates of future energy use worldwide are staggering.

On the other side of the coin, there is also bad news when one considers achieving progress in the world if transportation itself is severely constrained. At first glance, traveling less or reducing freight movements might not seem so bad. But remember the two-step dance, the ways transportation enables so much of what we do. The bad news comes from constraining the things transportation enables. That point seems not to be recognized by those critics who view transportation as a frivolous endeavor.

So we do use lots of energy—more than we could have imagined 50 years ago—and the worldwide use will almost surely continue to grow. The task ahead is to use it as wisely as we know how and to try to foresee and forestall major dysfunctions. Transportation is center stage in all these considerations, and our objective here is to illuminate a bit more this aspect of the issue.

Transportation energy consumption by mode

Table 13.1 tells us the energy use by each mode of transportation in the United States. As we said earlier, emissions are tied to energy consumption, so the energy consumption of each mode provides a good approximation of the contribution of the modes to pollution. It is not a perfect correlation because of modal differences in fuel mix (gasoline versus diesel fuel); differential application of emission standards to types of vehicles; and different operating cycles of warm up, acceleration and deceleration, idle, and cool down.

From a gulping fuel point of view, autos and trucks are the bad guys: about 80% of fuel use is by automobiles and light trucks (about 60%) and larger trucks (about 16%). (A light truck is a two-axle, four-tire vehicle; and many of these vehicles are used for auto-like passenger transportation.) For this reason, it is no wonder that some people attack automobiles and highways as the environment and energy evil. Coupled with the frivolous use notion, the attack is no surprise.

There has been big improvement. Almost 20 years ago, the federal government placed Corporate Fuel Economy Standards (CAFE) on the fleets purchased by consumers. In 1978 the standard was 18 miles per

Table 13.1
Percentage Energy Used by Transportation Mode

Transportation Mode	Energy Used (%)
Automobiles	39.1
Buses	0.8
Light trucks	19.9
Other trucks	15.9
Air	8.7
Water	6.0
Pipeline	4.0
Freight rail	2.3
Transit rail	0.2
Commuter rail	0.1
Intercity rail	0.1
Construction and agriculture	3.0

gallon (mpg), and since 1990 it has been 27 mpg. The requirement for light trucks reached 20.6 mpg in 1995. That worked, along with other things—perhaps including vehicle owners' perceptions of higher real fuel prices. In spite of the increases in numbers of vehicles and passenger miles of travel, fuel consumption by automobiles increased only about 8% when vehicle miles of travel increased by about 70%.

Tables 13.2 and 13.3 address the extent to which the highway mode's gulping of fuel results from there being so many cars and trucks or because of their being relatively inefficient. For instance, the efficiency of automobiles and rail transit is about the same (Table 13.2); but when overall consumption is considered, automobiles gulp more because they are used more (Table 13.1).

Table 13.2
Btu Per Passenger Mile, 1998

Transportation Mode	Btu
Automobiles	3,642
Passenger trucks	5,187
Motorcycles	1,784
Transit buses	4,452
Intercity buses	949
Commercial air	5,048
General aviation	9,825
Intercity rail	2,351
Rail transit	3,825
Commuter rail	2,889

Table 13.3
Btu Per Ton Mile

Transportation Mode	Btu
Truck	2,827
Large railroads	388

The energy used per passenger mile depends on the number of passengers, of course. For the automobile, for example, it is assumed that 1.6 persons are in the vehicle. The average auto with only a driver consumes almost 6,000 Btu per mile, but load that vehicle with a family of five and consumption drops to almost 1,000. A crowded transit bus is very fuel efficient, but the "on average" calculation has to include many miles of lightly loaded operations.

Recognizing the difficulty of achieving improved mileage as traffic now moves, modal switching is often advocated. Looking again at Table 13.3, we see that if cargo is shifted from truck to rail, then energy should be saved. Table 13.2 offers a larger set of options for switching—say, from general aviation or commercial air to intercity rail or from rail transit or transit bus to the automobile. But thinking a bit will dampen enthusiasms for mode switching. The modes are in differing market niches. Asking rail, for example, to take on the short-distance hauls that trucks serve suggests a 2,000-hp locomotive pushing a car or so here and there. That raises cost and service flags, and one wonders whether energy would really be saved.

Pollution

The pollution situation has similarities to the energy situation. Similar to the case in energy, pollution has multiple sources—transportation is not the only sinner. Limiting concern to transportation, highway vehicles are the worse sinners (Table 13.4).

Table 13.4
Percent of National Emissions

Pollutant	Highway Vehicles	Total Transportation
Carbon monoxide	62	77
Oxides of nitrogen	32	11
Volatile organic compounds	27	37
Particulate matter less than 10m	1	2
Sulfur dioxide	1	3

Also similar to the energy situation, there has been a reduction of emissions in some cases—especially in lead emissions, which have decreased by a factor of 90. Overall, emission growth has been curtailed during a period of increases in transportation activity (Table 13.5). Carbon dioxide is measured in tons of carbon, and its emission has slowed as the growth in the use of fuels has slowed.

Highway vehicle miles of travel roughly doubled between 1970 and 1994, while emissions decreased or remained about the same. Volatile organic compounds and carbon monoxide were down a bit; oxides of nitrogen and particulate matter remained about the same. Again, this is a good track record considering the growth in total usage over the period.

This has largely been achieved by cleaning the exhaust of automobiles. New autos are very clean compared to old ones, and the emissions of the fleet of new vehicles has gained dramatically in comparison to other sources of pollutants. For example, it is pointed out that the carbon monoxide emissions from operating a snow blower for an hour in Chicago are about the same as those from driving a new car from Chicago to Denver; the Chicago–to–Kansas City trip compares for volatile hydrocarbons. Relatively and absolutely, cars have gotten very clean.

Where are we going?

Where we are going depends on who "we" is; it is a we (already pretty much automobilized part of the world) versus they (the other folks) thing. In the United States the growth of auto-based mobility has tapered,

Table 13.5
National Transportation Emissions, 1970 and 1994 (million short tons)

Pollutant	1970	1994
Carbon monoxide	99	77
Oxides of nitrogen	9	11
Volatile organic compounds	15	9
Particulate matter less than 10m	1	1

the wars over modal market shares are pretty much over, and patterns of vehicle use are largely fixed. Those observations lead one to conclude that while the growth of mobility caused past increases in energy use, future demand-driven energy use and emissions in the United States will tie mainly to population growth. How long will the natural expansion of the population continue (excess of births over deaths)? What is the future of immigration? We cannot answer these demand-side questions.

Balancing demand and conservation and pollution control actions, the International Energy Agency highlights how increases in energy uses have been driven by mobility increases and estimates that OECD nations' carbon dioxide emissions and energy use will continue to grow through year 2010, a not unexpected finding. (The OECD nations are mainly well-developed nations.)

Looking worldwide at our energy-use crystal ball, it explodes, because energy use and emissions will increase vastly as the developing nations increase the income and wealth of their citizens, thus permitting them to turn to automobile use.

What can we do about it?

That's pressure for growth on the demand side. What about responding by improving vehicles? Mandated to improve efficiency, automobile manufacturers did the easy things, the things they knew how to do and that were most cost effective. The array of actions included reducing weight, air resistance, and rolling resistance. They also included more efficient combustion and drive train improvements that reduced friction and improved matching engine output to load. The list is longer, of course, and research and development continues at a high rate. A recent annual energy conversion conference, for example, saw a high of 416 technical papers [7]. Continued increases in work are to be expected.

The energy or environmental expert may wish that we would go beyond saying that there are lots of vehicle engineering work and review alternative fuels, electric vehicles, traffic management, road pricing, and fuel taxes. There is a flood of writing on those topics, and we will address some of them. But first we want to emphasize a very important point.

Everything we do involves tradeoffs

There is no free lunch for there are technical and political conflicts among achieving safety, energy, and clean air goals [8, 9]. The CAFE standards that raised the average miles per gallon of the U.S. car fleet has contributed to an increase in highway deaths because one of the primary techniques for improving fuel economy is to make cars smaller and lighter, which seems eminently sensible: why have a 3,000-lb car haul people who usually weigh in total less than 300 lbs? The downside is that smaller and lighter cars come off as distinct losers in collisions with larger vehicles. For this reason, public and private safety agencies caution against stiffer CAFE standards. Fuel-efficient, high-compression engines are prone to emissions of oxides of nitrogen.

Although there have been market distortions, CAFE has done what it was designed to do. Since the imposition of fuel-efficiency standards, the standard for automobiles has been increased by 52.8% and that for light trucks by 17.7%. (As we said, a light truck is any four-wheeled vehicle that is not a car and that weighs less than 8,500 lbs. Sports utility vehicles and passenger vans are classified as light trucks.) But what has happened to vehicle weight? The average car is about 500 lbs lighter today than it was 20 years ago, but the light truck has increased in weight by about the same amount.

This trade between fuel economy and safety was not widely recognized when CAFE was implemented. Had it been, it's very possible that the CAFE or its equivalent might have been structured differently, although it is not obvious to us how this might have been done.

The energy required to push vehicles through the air increases roughly as the square of the speed. The forces involved differ from mode to mode, but, speaking very roughly, at over about 30 mph overcoming air resistance becomes a significant energy user for cars, trucks, trains, and planes and there are big increases of energy use with further increases in velocity. Streamlining has lowered these forces and brought efficiencies, but further reductions in drag will come hard.

So we can improve fuel economy by going slower. But the productivity of all transportation vehicles and their crews varies directly with their average speed—the more miles they cover in a day, the more "productive" they are. For commercial vehicles, lower productivity reflects directly into higher costs, which in turn affect the cost of every good or service in the economy that uses transportation anywhere in its production or distribution. We don't think in productivity terms for private cars; here we see lost time and a lot of frustrated itchy feet. There is another trade involved: increasing speed to increase productivity and decrease lost time also hurts safety.

Muscle cars hurt fuel consumption. Energy is required to accelerate the mass of the vehicle, and faster acceleration requires more energy. Do away with muscle cars? Lots of people would protest. And many argue that the availability of a certain amount of muscle improves safety, like when trying to pass that truck on a two-lane road.

For lots of other techniques the cost is that of more sophisticated technology and increased complexity. One could seek strong, lightweight materials, but that is costly. Hybrid propulsion systems (electric motor and batteries plus internal combustion engine) would aid in recovering energy during deceleration. While such technological improvement costs have a way of coming down rapidly over time, they are still a reminder that there is no free lunch.

Changes are not costless and they, along with changes mandated by environmental regulation, have increased the real cost of the automobile. A full-sized Ford in 1950 cost about $2,000; and by 1998 its cost had increased to over $20,000, still a big increase even after the inflation factor of five or six since 1950. The reason is that consumers are buying more car per car. Air conditioning and front-wheel drive, for example, were not common in 1950 purchases. Without being exact we may observe that the kinds of product/price improvements that we take for granted for computers and many other electronic products are not characteristic of today's automobile. The technology is aged and locked in, it is a mature technology, and when mandates are laid on, it is hard to avoid increased real costs.

We have already made the point that whatever we do to cut petroleum consumption involves trades that are legitimate and that these trades will be made differently as the relative price or cost of various elements in the trades shift over time. It is unlikely that there will be arbitrary, wholesale curtailment of any feature or performance characteristic that is not driven by economics.

Higher fuel prices?

The price of a gallon of fuel in Sweden and West Germany is about four times that in the United States. Some say, "If gasoline prices were the same as in Europe (or Japan), then...." One cannot disagree, for if price goes up consumption decreases.

The situations are not strictly compatible because of the less-dense settlement pattern in the United States. The average distance an auto is driven in the United States per year is about one-fourth more than that in the comparison nations. A possible reduction in travel is only part of the story, for with increased fuel prices one would suppose that purchasers would elect more fuel-efficient vehicles. But remember our tradeoffs: that may not be a desirable choice when the safety implications of lighter–smaller are recognized.

What is politically practicable looms large in the United States and based only on a hunch we imagine that the economist's favorite of higher fuel taxes is resisted by concerns about how the government would spend the tax collected. How high would taxes have to be to make a difference is also a question. In 1995, Nivola and Crandall estimated that quadrupling the then–federal tax to about $.50 would decrease U.S. automobile fuel consumption by about 12%, but that would decrease world petroleum consumption by less than 1% [10].

The use of alternative fuels is being promoted. Natural gas (mostly methane) is inexpensive and in good supply, so compressed natural gas or natural gas converted to methanol are favored fuels. But the costs of compression or conversion reduce the cost advantage of natural gas. Its big advantage is not that it might decrease hydrocarbon consumption, but that it is cleaner than gasoline.

Ethanol made from vegetation is favored by the farm lobby. Ethanol is relatively expensive and producing-collecting raw materials for production consumes much energy. It has the advantage of being made from a renewable source.

Hydrogen is sometimes proposed for use in internal combustion engines. But the cheapest source of hydrogen is from petroleum products. Further, there are cost and weight disadvantages (fewer British thermal units per pound than gasoline) that translate into reduced driving ranges and other problems. But there is another possibility.

The fuel cell and the hydrogen possibility

The fuel cell of interest here is a device that produces electricity by combining hydrogen with the oxygen in the air. Thus, it has the same advantage as the gasoline engine: it doesn't have to carry the weight of all its fuel on board the vehicle because it gets the oxygen required almost for free from the air. (The storage battery, on the other hand, has to carry all the weight of the chemicals storing the energy on board the vehicle—giving it about a 5:1 weight disadvantage in competing with gasoline for vehicle performance.)

Many of the automobile companies have very serious programs to develop the hydrogen-air fuel cell for vehicular use, and the mood is optimistic that they will be successful. Technical problems are associated with the fuel cell stack, fuel processing, and ancillary devices [11]. The impact they will have on the energy and emissions picture, however, is cloudy. Requiring high capital investment, they might first be used in ships, trucks, and locomotives.

If the hydrogen required is produced through some external process that doesn't itself produce nonbenign emissions, then the only emissions will be those from the vehicles themselves and they will be nothing but water vapor (the combination of the hydrogen with the oxygen from the air).

This is not what the automobile companies are counting on; the plan is to produce the hydrogen on board the vehicle from some hydrocarbon compound such as methane. This not only complicates the on-board system, but it implies that the conversion will produce carbon dioxide and possibly other gases as a byproduct. The level of emissions are expected to be low, but not negligible.

But if by some chance there is a major attitude change toward nuclear energy generation (use electricity for electrolysis of water to separate hydrogen and oxygen), or very large scale solar production of hydrogen becomes feasible, or if some other process is developed, then hydrogen holds the promise of breaking transportation's strangle hold on the world's oil supply. It would be a wrenching change in the world's energy industry, but it would undoubtedly be taken as many small steps over many decades. The phrase "creative redirection" would be far more appropriate than "creative destruction." .

This hydrogen substitution evolution is the only path that we can see now that could remove the specter of ever-increasing petroleum consumption by transportation. We add, however, that the world is full of surprises for us all.

A parting thought: wouldn't it be ironic if we end up producing so much water vapor and new cloud cover that the world is threatened by global cooling?

What, then, to do now?

There are many things that could be done to slow the increase in petroleum consumption and concomitantly its side effects; we've talked about many of them. Manufacturers could make increased use of plastics, aluminum, and composites to reduce weight still more (but don't forget our trades). New energy-conversion recipes might include the use of turbines running at constant velocity coupled to flywheel energy storage. Proponents combine options and talk of 100-mpg vehicles. The list of could-do things is long—as thumbing through technology enthusiasts' magazines will reveal.

But there are reasons why the options have not been introduced: they either cost more (new materials, new combustion technologies), reduce product quality (smaller and or lower performance cars), or raise safety issues (lighter weight vehicles). It is not surprising that the 27-mpg CAFE standard has stuck and that suggestions to tighten it meet strong objections. It's those darn tradeoffs again.

So given all these tradeoffs, and assuming a perfectly rational world, is there anyplace else to go? We have done the easy things—picked the "low-hanging fruit." To mix our metaphors, there may still be blood in some long-range turnip, but we don't immediately see it in the one we have in front of us.

For pollutant emissions, we have also pretty much picked the tree bare, we have already reached for the fruit that is inexpensive to grab, and more may not be worth getting. For example, even if the zero-emission (probably electric) vehicle for California comes to pass, it will hardly make any difference because the new cars coming along are already so clean.

But we have to improve; we aren't there yet. Still, our thought now is that inaction might very well be the best action. New options are coming down the road, in technology and perhaps in people's attitudes toward readjustment. A pause now to think about these possibilities and to carefully weigh them against any downsides is probably our most fruitful path.

All is not lost. Look at the way the world works. Growing scarcity breeds higher prices, creating incentives for alternatives. Options to respond to the changing conditions are exercised, and the world rolls on. It has not, in the long run, caused more-expensive and less-efficient transportation. We are, in fact, counting on this process observed over and over in our past to keep us out of trouble. We have more technological responses at our command than ever before in history.

References

[1] Davis, Stacy C., *Transportation Energy Data Book: Edition 18*, Oak Ridge, TE: Oak Ridge National Laboratory, 1998.

[2] Bureau of Transportation Statistics, *Transportation Statistics Annual Report 1998*, Washington, DC: U.S. Department of Transportation, 1998.

[3] Marsh, G., *The Earth as Modified by Human Action*, New York: Charles Scribner, 1898. (Revision of *Man and Nature*, 1863.)

[4] Altshuler, A., et al., *The Future of the Automobile*, Cambridge, MA: the MIT Press, 1984.

[5] Brown, L. R., C. Flavin, and C. Norman, *Running on Empty: The Future of the Automobile in an Oil Short World*, New York: Norton, 1979.

[6] Jevons, W. S., *The Coal Question: An Enquiry Concerning the Progress of the Nation, and the Probable Exhaustion of Our Coal-mines*, 2nd ed., rev., London: MacMillan, 1866.

[7] *31st Intersociety Energy Conversion Conference Proceedings 1996*, Warrendale, PA: SAE, 1996.

[8] Crandall, R. W., et al., *Regulating the Automobile*, Washington, DC: Brookings Institution Press, 1986.

[9] Bryner, G. C., *Blue Skies, Green Politics: The Clean Air Act of 1990*, Washington, DC: CQ Press, 1995.

[10] Nivola, P. S., and R. W. Crandall, *The Extra Mile: Rethinking Energy Policy for Automotive Transportation*, Washington, DC: Brookings Institution Press, 1995.

[11] *Fuel Cell Power for Transportation*, papers from the 1999 International Congress and Exposition, Warrendale, PA: SAE, 1999.

Part IV

Between Cities: Rail and Other Ground Transportation Systems

14

The Innovation that Changed the World: How Did It Happen?

The beginning of the era of change, wherein, one after another, some new invention, method or situation compels a fresh adjustment of behavior and ideas.

Durant, *The Lessons of History*

We began this book with what many folks would classify as some fairly bold leaps: first the car that can drive itself, and then the almost heretical thought that congestion does not fall in the same category as death and taxes. Then we raised the possibility that there may be an entirely new high-density, carless variant enclave to add to our already variegated cities and perhaps some new kinds of neighborhoods. We will talk in

167

future chapters about other things: innovation in the railroads and some altogether new systems for intercity travel and freight movement.

The operative concept throughout has been change—innovation, evolution. In this chapter we'd like to broaden our perspective of this process of change itself and our current place in it.

We will make a brief trip back to the England of 1800, where one of the most important transportation developments of the last 300 years was beginning to take shape. The specific elements in the drama played out then have long left the stage, but the plot lives on—as we shall see.

England in 1800—"Built out"

Most Englishmen 200 or so years ago would have said that their transportation system was pretty good. Service was available nearly everywhere, provided by earlier waves of expansion of harbors, river improvements, canals, tramways, parish and toll roads, and a thriving sailing industry providing coastal shipping. The evolution of these modes hadn't stopped—there was a continuing search for marginal improvements—but these modes were mature in the context of the technology to which the world was accustomed.

Tramways fed freight to rivers or canals wherever the economics warranted. Passenger trams would soon begin to be used in large cities. Trams are nothing more than wagons on rails moved by men or horses or gravity, a primitive forebearer of the railroad car. Almost a century later they would transition to electric streetcars, which the British continued to call trams.

Road improvements were also in place, with most of the through-roads being toll roads. Before and after the turn of the century, Metcalfe, Telford, and McAdam strived to improve toll road construction, maintenance, and management. Ways were sought to extend roads into remote highlands, and, as is done today, the national government gave money to match the funding provided by local governments.

Intercity passengers had the choice of walking, horseback riding, and either "fast" coach or slower wagon services. They could travel by canal and take advantage of fly boats (fast boats) or use ordinary services. Freight movements had a similar set of alternative services. Bulky freight was moved to and from canals in wagons. There were rough spots: improved management and more rational pricing were needed. But by and large, contemporaries would say that satisfactory transportation systems were in place, and plans and actions were in progress to improve their workings.

Parliamentary debates centered on such matters as franchises for toll roads and the effectiveness of toll road companies. There were efforts to tie the level of tolls to damage to roads. Issues such as the appropriateness of tolls for those going to church on Sunday and the control of the evasion of tolls were also given attention. Indeed, it is said that during the years of the American Revolution, Parliament spent more time debating road policy than it did discontent in the colonies.

The transportation system in 1800 satisfied its English users in the sense that its performance was an everyday fact, and peoples' activities and expectations were geared to that which was available.

The picture in 1800 is thus one of an England "built out" in the transportation technology of the earlier century. Few even imagined that things could be substantially improved or paused to think how life and commerce would change if they were.

But change they did. The breaking away from this make-do-with and improve-what's-there attitude occurred as ways were sought to develop the Southern Durham coal fields in Northeast England. The problem to be solved was that of moving coal from mines in the vicinity of Darlington 25 miles to Stockton-on-Tees for coastal shipment to London, a great English coal market of the times.

There were rival schemes: improved roads, canal improvements along the Tees River, and tramway service. Not unexpectedly, considering that modes had already been developed where practicable, each scheme had its ills. Road wagon haulage was too expensive, the tramway would have to be a very long one, and canals would have to overcome the heights with many expensive canal locks.

The Stockton and Darlington Railway

Local landowner Edmund Pease was persuaded that a tramway would serve best, and he obtained Parliamentary sanction for a tramway in

1821. An engineer named George Stephenson, an expert in coal mine steam equipment, heard of Pease's scheme, visited Pease, and was hired as an engineer for the tramway route.

At some point Stephenson persuaded Pease that steam locomotives might be superior to horses for pulling wagons, in spite of the fact that accounting studies had shown that they gave no marked advantage. Pease returned to Parliament in 1823 and obtained sanction for the use of locomotives. Construction proceeded, and on the 27th of September 1825 the railroad era was announced when a steam locomotive pulled a train from Darlington to Stockton. The rest, as we say, is history.

It is said that experience is a hard teacher because the lesson comes after the experience. That is true, but we would say that experience may be an ineffective teacher because lessons are so often ignored. But the Stockton and Darlington track not only drew heavily on prior lessons, borrowing liberally from earlier roadway and canal experiences, but going beyond just cheaper and faster to discovering completely new ways to do things.

There are many lessons to learn from the Pease, Stephenson, Stockton, and Darlington story. One is that the conventional wisdom that everything that could be done has been done is often incorrect. Another is that new services and markets are often discovered by what looks like an accident; foresight is nice, but fallible.

At the time, and as is often true today, technical experts, managers, and politicians viewed the available ways of producing services as the only possible ways. That was conventional wisdom. Improvements were sought by refining technologies, policies, and institutions and by extending services to their practical economic and geographic limits. No one imagined doing something new.

Pease needed all the performance improvements he could find. Political actions by Member of Parliament Lambton had imposed low maximum haulage charges for coal to be shipped from Stockton to London. This low charge was expected to make coal movements for the London market uneconomical and to limit Pease's coal shipments to local markets where haulage charges were not so constrained. This would preserve the London market for Lambton's constituents.

Pressed hard to keep costs low, the Stockton and Darlington railway was built with careful attention to cuts and fills to limit the need for moving lots of dirt. In places, and like other tramways, it used cables to pull the trams to avoid the need for cutting into hills. The result was a well-engineered, state-of-the-art tramway.

But there was also discovery, the finding of new ways to carry out old functions. Stephenson had experience with locomotive construction and operations while at the Killingworth Colliery, and he had pioneered with others in their development. Stephenson-built locomotives had been in use at Killingworth, and locomotives by other builders were in use at other collieries for private haulage.

It was not at all clear at the time that the substitution of locomotives for horses would improve performance. But Pease—Pease the entrepreneur, Pease the man who put up the money—made the decision to order a pair of locomotives be built (even though horses hauling carts were still expected to provide most of the service). Pease's willingness to try locomotives was the first step in discovery.

Discovery continued after the route was opened. Modeled on canal operations, Pease initiated services using driver-owned horse-drawn wagons paying tolls for the use of facilities. Although passing sidings were provided, the facility tended to operate at the low end of the speeds of horse-drawn wagons, from 2 to 4 mph, with some wagons overtaking and some waiting at sidings.

It was soon discovered that in this situation locomotives did, in fact, have advantages compared to horses. Locomotives could operate at from 10 to 15 mph and, because much of the track when wagons were loaded was downhill, pull 12 or more wagons. This seems not to have been anticipated, for early uses of locomotives by mine operators had not suggested the advantages found when locomotives were placed in line haul service (service with reasonably long stretches between stops).

With the discovery of these advantages, the operations protocol was changed as Pease and Stephenson added more locomotives. Soon 500,000 tons of coal were being transshipped annually at Stockton for the London market. By the 1840s the Stockton and Darlington was a heavy-haul route (although the typical heavy-haul route of today moves fifty times that tonnage).

There was even more discovery. While passenger services had not been much imagined, passengers had ridden in specially rigged wagons and empty coal wagons on the first Darlington to Stockton train, the opening day train. This uncovered the latent demand for passenger service, and independent operators began to offer horse-drawn, scheduled coach service. It was natural to offer train passenger service also. As higher speed locomotive-train service emerged, passenger services were further improved.

Pease had thus discovered—innovated—locomotive-hauled trains and the organization to operate them and expand the service. His railroad not only provided the low-cost freight services he set out to produce but passenger services as well. It wasn't many years until the temperance preacher Thomas Cook discovered market response to his package tours for working people. The travel industry was discovered. Discovery everywhere.

The recipe for discovery

What was the recipe for discovery, for breaking away? It had a design essence, a new design laid the base for a new departure. The design ingredients weren't new. Stephenson's route planning and construction involved straightforward borrowing of existing techniques, mainly from the earlier experiences of others with building roads and canals. Much was borrowed from tramway technology: cable haulage using stationary steam engines where there were steep grades, track support, rails, and track gauge.

The plan to allow individuals to operate horse-drawn wagons after paying toll charges was borrowed more from toll roads and canals than tramways because most tramways of the times were owned and operated by mine properties. Financing, different tariffs for classes of commodities, and institutional arrangements were modeled on the canal pattern. And as the railroad form emerged, rules for managing labor were borrowed from the rule books of manufacturers.

Stephenson's locomotive wasn't new either. A man named Richard Trevithick had built the first one some 22 years before the Stockton and Darlington.

Trevithick had built several steam carriages in 1796 to operate on roads. (This is over 100 years after Newcomen's and Papin's original notions: the steam engine had come a long way from that seminal step.)

But the roads of the day were too rough for the machinery, and Mr. Trevithick finally turned his attention back to his main business of building stationary high-pressure steam engines, mostly for use at mines.

In 1803 he was building an engine for an iron rolling mill in south Wales. The story goes that the ironworks owner, Mr. Samuel Homfray, had great faith in Trevithick's high-pressure engines (2 to 3 times atmospheric!). He also apparently had little tolerance for the skepticism of a neighboring mill owner and seemed to have talked himself into a 500 guinea bet that Trevithick could build a steam locomotive that could pull a 10-ton load at 5 mph.

Mr. Trevithick's engine won the bet, pulling not only the 10-ton load but allegedly the additional weight of the some 70 people who jumped on board for the ride. The only problem was that the weight of the train broke too many of the cast iron rails, and so the locomotive was retired.

Trevithick built at least two more locomotives: the last, four years later, reached 12 mph—almost five times faster than a horse could pull a canal boat. They were sophisticated devices, given the state of the art at the time.

This story of Trevithick's locomotives has a familiar ring: they worked quite well, but they couldn't attract financial backing; horses were still thought to be cheaper. Richard Trevithick gave up building devices nobody seemed to want and once again turned his full attention back to his primary trade of building stationary engines.

So much for being first technically when the market niche, where it might be successfully applied, had not yet come into focus; this is the reason we have Pease rather than Stephenson in the forefront of our story.

(As an aside, while Mr. Trevithick was inventing the locomotive, the groundwork was being laid in America for the largest locomotive market the world would ever see: President Jefferson, having just made the Louisiana Purchase and believing the Missouri River could connect to rivers leading to the Pacific, thus providing a water route from coast to coast, was dispatching Captain Merriwether Lewis and Captain William Clark to explore this route. The water route wasn't there—the Rocky Mountains got in the way—but their exploration gave the first hint that the new nation might extend beyond Ohio.)

After Trevithick, other machines had been built by Blenkinsop, Hedley, and Stephenson himself to operate at various mines. Indeed, the ingredients of Pease's railway were state of the art at best, and the result was so tramway-like that many critics dismissed it. Very little invention was needed. The secret was combining old ingredients in new ways. And having the nerve to go ahead.

Nothing is perfect, and mistakes were made. With hindsight, the use of cable haulage over steep grades and early involvement with independent wagon operators were mistakes. Some things were done poorly. The first carriage built specially for passenger hauling was so heavy that it could not easily be pulled upgrade from Stockton. And even with experience, many lessons were yet to be learned.

Stephenson was soon involved in other railroad endeavors. These were built mainly with freight service in mind, yet passenger service was the first market to build quickly. That was the case, for instance, with the Liverpool Manchester Railway built to link manufacturing in Manchester with the docks at Liverpool. Actually, Stephenson may have been aware of passenger service potentials but did not discuss them in order to avoid conflicts with road coach operators. At any rate, coach operators seemed not to have learned the rail passenger service lesson, for unlike canal operators they did not see early railroads as a threat to their business.

The Stockton and Darlington railway succeeded not only because it was technically successful but because it operated in a market niche where the need was real. The route was selected to serve this need. To say that success in a market niche is necessary for successful innovation borders on stating the obvious, yet it needs to be stressed. It is clearly not obvious to everyone, or there would be less failures on that dimension.

The market is where hype meets reality. Success in the initial niche is necessary because it provides the stage for learning how to refine services, enhance services, and discover new services. And being first in the niche leaves its stamp, sometimes seemingly forever, as the next part of this early story illustrates.

Stephenson built the Stockton and Darlington railway with a 4-ft, 8-in gauge; he added the extra half inch when he built the Liverpool and Manchester. (Truly hairsplitting optimization!) This magic gauge was inherited from the size of the horse-drawn trams already being used to move coal from the mines to the canals.

A few years later the famous bridge engineer, Isambard Kingdom Brunel, wanted to go faster. He used bigger wheels, since for every revolution the wheel rim travels further on a bigger wheel. This raised the center of gravity of the cars, making them a bit tippy for the narrow gauge rails. So he increased the distance between the rails—the gauge—to 7 ft. He found funding to build the Great Western Railroad from Bristol to London, where by 1840 he was routinely operating his trains at 60 mph.

Mr. Brunel was the man who had already designed and built a very large steamship, the 233-ft Great Western, which came within four days in 1838 of being the first ship to cross the Atlantic entirely on steam power. The Great Western was, in fact, the first true transatlantic ocean liner. He also designed and built the huge, 680-ft-long Great Eastern steamship, intended for nonstop service to Australia around Africa. Launched in 1851, it was just too much for the market or for the technology and was a commercial failure. No one has ever accused Mr. Brunel of thinking small.

Back to railroads. In 1844 a committee in England convened to examine the issue of gauge and what should it be. The committee decided that 4 ft, 8½ in was indeed too narrow, but they also decided that so much track had already been built that way that it was too late to change. On their recommendation the Parliament passed the Gauge Act of 1846 decreeing 4 ft, 8½ in. Only 16 years after the Stockton and Darlington, it was already too late to change!

We are still building track 4 ft, 8½ in apart, the tyranny of an almost ancient technical legacy.

But to summarize: the recipe for discovery used familiar ingredients borrowed from here and there. Some ingredients, such as operations by independent coach operators, were rejected as the railroad emerged. The qualities of most ingredients, such as the locomotive, were improved with experience after a moderately successful design was found.

The ingredients were not new, it was the way building blocks were put together, the system design, that was new. The system design involved both performance improvement and functional refinement, and the discovery was the railroad and its services in infant form. Pease gave a glimpse of what the railroad could do, and Stephenson and Brunel in England and John Stevens in the United States, among many other leaders, went on to refine the technology and the market.

The design incorporated both hard and soft technologies. Hard technologies included rails, bridges, and locomotives. Soft technology refers to ways of doing things, such as pricing services, obtaining financing, and the organizational arrangements for managing work.

Finally, success in a market niche was followed by refinement and expansion of ways to provide services. More broadly, there was the discovery of a broad array of functions supporting economic and social development.

The Pease story tells one birthing experience, a well-documented case. Just 50 years after service was begun, S. J. Jeans in 1875 provided a look back at the Stockton and Darlington in his *Jubilee Memorial of the Railway System: A History of the Stockton and Darlington and a Record of its Results* [1]. He provides a tribute to Edmund Pease; we have also placed Pease front and center in the story. That's not to diminish the importance of Stephenson who also deserves fame as an innovator and who went on to become a famous railroad builder. It is proper to recognize Stephenson as the railroad inventor, as most do. But the vision and entrepreneurship of Pease played a critical role, and we have spotlighted him to underscore the importance of such innovation managers.

The British transportation history story is exceptionally well documented. Our favorites include books by Ranson [2] and Dyos and Aldcroft [3].

We go on to part two of our story in the next chapter.

References

[1] Jeans, S. J., *Jubilee Memorial of the Railway System: A History of the Stockton and Darlington and a Record of its Results*, London: Longmans, Green, and Co., 1875.

[2] Ransom, P. J. G., *The Archaeology of the Transport Revolution, 1750–1850*, Tadworth: World's Work, 1984.

[3] Dyos, H. J., and D. H. Aldcroft, *British Transport: An Economic Survey From the Seventeenth Century to the Twentieth*, Leicester: Leicester University Press, 1971.

15

America and the Two-Step Dance

Like ripples in a pond…

The birth of the railroads did far more than just improve transportation, it was a key development in energizing the whole Industrial Revolution. In *The Role of Transportation in the Industrial Revolution*, Rick Szostak attributes England's lead in this historic metamorphosis to their central role in inventing and implementing this new technology of rail [1].

Then, as now, ideas were light baggage and traveled easily, and news of Stephenson's railroads quickly ignited interest in both Europe and America. Especially in America, an expanding country still short of decent roads, there was enthusiasm for short routes joining cities on the eastern seaboard and longer routes penetrating from the seaboard into the interior. A railroad from Charleston leading inland into cotton-farming country and another from Baltimore reaching toward the lake states were soon under development.

But English railroads did not transplant easily. It quickly became clear that English practices did not fit well with American circumstances. Britain was a densely settled, compact country; and it often took, literally, an Act of Parliament and a lot of money to get a right-of-way. The British attitude was that once they had a right-of-way they were going to run lots of traffic over it forever and hopefully at higher and higher speeds. So they were willing to make large investments in cutting through hills, building bridges, and boring tunnels to produce straight track with very low gradients and only gentle curves.

In America there was more space and less money. It was easier and cheaper to follow the terrain than to dig through it, so there were sharper curves and steeper gradients. Track was less substantial than English track: iron and steel were mostly imported and therefore expensive, routes were rapidly getting longer, and capital was always short. Some English locomotives were tried but lacked power for the steeper grades, didn't track well on the sharper curves, and were too heavy for the track. Further, the English locomotives burned coal, and America needed wood-burning locomotives—the Appalachian coal fields were yet to be developed.

John Stevens and others rose to the challenge, and it wasn't long until locomotives were being designed and built in America for operation on American track. The key innovations were fire boxes suitable for wood, bogies (swiveling "trucks"—wheel sets) that aided both locomotives and cars in going around curves, and track structure using wooden ties. An innovator with the charming name of Isaac Dripps developed the cow catcher, a useful device in an era when locomotives were small and cows ran free more often than not.

The railroads in the United States first were built on a stage where settlement patterns had already been determined by water transportation. Cities had emerged at the heads of navigation on the bays and rivers of the eastern seaboard; this was where commodities were transshipped from land to ship. Those cities that had waterpower available began to serve as milling and manufacturing centers: the textile mills of New England, for example.

With access to the Gulf, New Orleans served commerce near the mouth of the Mississippi, and Cincinnati, St. Louis, and Pittsburgh were transshipment points for the Mississippi system trade. Indianapolis is the only sizable city in the eastern part of the nation not sited during waterway days.

This sketch of transportation and settlement could be extended to the cities of the Great Lakes and westward to the heads of navigation on the western rivers. Indeed, it could leap to San Francisco and ports to the north. Enough said; our purpose is served.

The railroads changed the comparative advantage of places and enabled new kinds of specialized activities. Adam Smith's ideas about the division of labor and market size applied [2]. Railroads expanded supply and market areas and enabled the specialization of production. Chicago became the nexus of rail travel from the east to the west, and Chicago's growth surged ahead of St. Louis's. Chicago soon became the grain merchant and hog butcher to the world because the railroad tied it to the fertile midwest as a source and to the east for a market [3].

The commercial superiority of one city or region over another depended to a very large degree on its strategic placement in this new network over which goods could move.

Mercantilism ruled as cities and states competed for railroad, inland waterway, and ocean-shipping facilities; they often supplied capital to aid facility development [4]. The pressure increased for public-sector financial support to provide railroad access nearly everywhere; similar pressure persisted even as later when the highway and air modes came along. The winners in this game usually prospered, and the growth of some regional centers accelerated. For the losers, growth stagnated.

Shallow stuff

Our observer and sometime critic, the Martian, had been listening intently, and he interrupted our narrative. He was typically blunt: *"It sounds like you are saying that transportation developments just move prosperity from one place to another. It creates competition about who gets the goodies. Sometimes it is cities wanting airport money because they think it is good for the merchant economy; sometimes it is folks wanting the jobs that spending money can bring. All through this book you are sketching how transportation might be improved. It sounds like you are saying that improving it will just move money from one person's pocket to another's faster! Why bother!"*

His friend, the Venusian, sometimes had a bit broader perspective: *"Look, you've already told us enough to see that there is more to this than just competition for pieces of the pie. The world has changed too much and the pie has grown too much to think that the only impact of better transportation was to let the Chicagos grow faster than the St. Louises. Give us a richer picture than your mercantilism-settlement pattern stories."*

The two-step dance

Let's underscore what we have said before by returning to our two-step dance metaphor. Remember from Chapter 12 that the first step of the dance is the substitution of cheaper, faster, or better service in an existing market niche.

The second step in the dance occurs when innovative folk think of ways to do new things using the newly created services—entirely new market niches are uncovered, or old ones significantly expanded and remolded [5, 6]. It is through the second step in the dance that transportation impacts the rest of the economy, creating opportunities for entirely new activities, or improvements in the way existing ones are carried out. We think that in the long run this second step has been and will continue to be the more important of the two.

Transportation itself is changed as the dance continues. Older modes either shrink or reorient their operations to focus on a different set of markets. The combination of air, personal auto, and the truck have changed the nature and pattern of rail service, eliminating passenger rail service in all but a few markets in this country. Trains and trucks and cars have virtually eliminated the horse-drawn vehicle. Transportation over inland waterways has not been reduced, but its market niche has been dramatically altered to primarily bulk commodities carried in barges. In each market niche the substitution is typically not complete, but the dominant mode changes.

These substitutions result in new patterns of transportation capability that, for any city or region, change not only the costs of access to old trading partners but make new ones available. The change in the available markets and new sources of supply induces a shift in the relative stress on the already ongoing activities in each. Comparative advantages do change. (We think the Martian overstated the case: we admit this is obvious, but deny that it's as shallow as he depicts it.)

Digging deeper

Since English railroads didn't fit the American scene, an industry was started in the United States to build American track and American trains.

This new rail and train manufacturing industry created the first big market for steel in the United States; and since importing British steel was expensive, it made the development of an American coal and steel industry almost inevitable. This new steel industry took a while to grow to large scale, however; it didn't happen until the Mesabi range was opened up as a source of ore. Of course, the trigger that opened the Mesabi was better transportation, both by rail and by ore steamers on the Great Lakes.

Railroads' operating necessities mothered invention and innovation in organization, finance, the art and science of construction, and later such technologies as the telegraph [7].

The transportation that the railroad furnished broadened potential markets, motivating and supporting whole new industries [8, 9]. We see the railroads opening farmland in the middle west. Looking more closely we see things in the middle west developing differently. Expanded market access justified larger farms and investment in mechanization; Virginia reapers mechanized grain harvesting. Scale encouraged the growth of grain and meat futures markets and marketing systems that reached across the country, even the world [3].

The downside was New England farmers hit hard by competition and the long-run fate of the small farmer; the upside was increased productivity leading to higher incomes for workers and increased choices for all consumers.

In contrast to waterways, railroads were not handicapped by either low water or by ice in the winter. Railroad services enabled year-round operations. Tied to markets by the telegraph and free from seasonal limits on shipping, large-scale production could be quite efficient; continuous steel production is a case in point.

As the railroad era went along, an enormous array of products was developed as specialization and scale economies increased efficiency and better ways to distribute products were devised. Consolidated meat-processing plants came along with the refrigerated freight cars and trucks that could move their output, and lumber mills—now able to sell into larger markets—produced standardized products.

The railroads filled one market "niche" and, in doing so, gave birth to a multitude of others, doing old things in new ways and doing new things.

The railroads, and later trucking, let the factories move to cheaper land so that the factory could be rearranged for greater productivity. Transit and the car freed the worker from having to live within walking distance of the factory and the factory owner from worrying about where his workers might live. The telephone and the telegraph reduced the need for physical contact to communicate and coordinate.

The steam engine made factories independent of running water for their primary source of power. (The spectacle of large leather belts transferring the power from the central power shafts to the individual work stations can be seen today in the Smithsonian exhibit of the world of 1876.) Later the small-scale electric motor let the factory spread out, no longer tied to the steam engines or water wheels that supplied the power.

So combining with other processes and activities, transportation gave us the new ways to do things that have shaped modern production (agriculture, mining, and industry generally) as well as distribution (warehousing, retail stores). Larger and growing markets increased the scale of operations and thus motivated and enabled greater specialization, increased productivity, and the development of new products. Sources of supply were broadened. Consumers had more choices and so there were consumption innovations as well as production and distribution innovations.

This metamorphosis of our economy started by the railroads was accelerated by the motor vehicle. The truck offered a new dimension to goods movement. The automobile broadened choices for attendance at schools, social events, theaters, and churches and in doing so reshaped many of our social patterns. It broadened horizons through dramatic increases in personal travel. The automobile set the stage for tourist courts, then motels, and recreation parks.

The modern, efficient supermarket largely replaced the neighborhood grocery store because the automobile lets the market draw its customers from a larger area and the truck permits it to be located anywhere the market exists. And scale expanded the varieties of fresh vegetables, meat, and baking goods available.

It has all happened in a hurry. By the mid-1920s, within two decades of the real take-off of the Model-T, roughly two-thirds of American families had a car. The time it takes for things to change in the future might be even shorter than we have thought, and impacts may be more far reaching than we imagine.

Over the sweep of the centuries we have evolved toward more diversity, from having only a few kinds of transportation services to having lots of kinds of services. Drawn on a piece of paper, this evolution looks like a tree. The trunk of the tree represents walking on paths and floating on rafts. Branches low on the trunk represent beasts of burden, horses and oxen and wagons, and sailing ships. More branches were added as time went, and a few died as they became obsolete. Most of the branches continued to grow so that each now represents many, many varieties of service.

Increases in specialization and in the diversity of choices available are probably the major social and economic trends, deep and long-running, trends that have encouraged an equivalent diversity in individual tastes in work, recreation, social life, and consumption. Folks tailor work, recreation, marriage, education, and worship in highly diverse ways, but in ways that suit their desires. To us this says we should strive to seek development directions for transportation that increase the number of choices available and that enable folk to specialize as they desire and to increase variety in their personal lives, and seek directions that increase productivity in our economy to provide the wherewithal to enjoy those choices.

Our Martian friend reconsiders: *"Now that is pretty heady stuff."*

But what we've described is only what yesterday's improvements have done for us. Although many analysts like to restrict the impacts of rail development to social savings [10, 11]—the savings because moving people and things was cheaper than before—transportation improvements can do much more. As economic historian Paul David points out, social savings is too limited a measure for economy-changing developments [12]. They may enable innovations, new ways of doing things, and new options for production and consumption. Today, that lesson often seems forgotten, the past seems out of sight and out of mind.

Creative destruction and the devil we know

There is always a loser in the contest between change and the status quo. Anything new plays devil and angel roles, and which is which is not always completely clear even in hindsight.

If change prevails, then old arrangements are made obsolete. History lessons tell us—and we repeat them here—how the development of the railroad brought agricultural products from the lake states to the markets of east coast cities. That new arrangement was disastrous for New England agriculture and, by providing an alternative to the river-gulf-ocean route through New Orleans to the east coast, it slowed the growth of that key city.

Streetcars and, later, cars opened land for expanded urban settlement and enabled the development of outlying shopping centers that competed with central business districts. The airplane reduced the markets for long-distance passenger trains. The container and ocean liner and airplane brought the world's goods to many new markets, displacing products that were there before. In the instantaneous judgment, the status quo lost in all these cases. Is the new status quo better than the old? Most of us think so, but it still depends on whom you ask.

As old sometimes-proponents of new systems and new ideas, the authors are acutely aware that almost no matter how flawed the status quo, the burden of proof is on the change, simply because change upsets too many apple carts. But if change is thwarted, then we never know.

United States in 2000: "Built out"

Analogous to England's transportation system in 1800, our transportation today largely satisfies us in the sense that its performance is an everyday fact and our activities and expectations are geared to what we have. Roads and highways let us go nearly everywhere. We have an extensive rail network that has actually been shrinking in miles of track (but not in freight traffic) as some parts are unable to compete with trucks. We have unprecedented air service and major freight movements on rivers and canals, and every city of some size has some form of urban transit.

There are rough spots, such as congestion, air quality problems, and energy consumption. Some say improved management and more rational pricing is needed. But by and large, contemporaries would say that satisfactory transportation systems are in place, and plans and actions are being taken to improve their workings. As was the case in England in 1800, U.S. transportation systems compare well with those in other nations.

The picture in 2000 is thus one of a United States "built out" in the transportation technology of the passing century. Few even imagine that things can be substantially improved or pause to think how life and commerce would be altered if they were.

References

[1] Szostak, R., *The Role of Transportation in the Industrial Revolution: A Comparison of England and France*, Montreal: McGill Queens University Press, 1993.

[2] Smith, A., *The Wealth of Nations*, Chicago, IL: The University of Chicago Press, 1976 edition of the original 1776 publication.

[3] Cronon, W., *Nature's Metropolis: Chicago and the Great West*, New York: W. W. Norton and Company, 1991.

[4] Dobbin, F., *Forging Industrial Policy: The United States, Britain, and France in the Railway Age*, London: Cambridge University Press, 1994.

[5] Garrison, W. L., and R. R. Souleyrette, "Relations Between Transportation and Production," *Transportation Research Record 1262*, Washington, DC: Transportation Research Board, National Research Council, 1990, pp. 21–30.

[6] Garrison, W. L., and R. R. Souleyrette, "Transportation Innovation, and Development: The Companion Innovation Hypothesis," *The Logistics and Transportation Review*, Vol. 32, 1956, pp. 5–37.

[7] Chandler, A. D., and S. Salisbury, The Railroads: Innovators in Modern Business Administration," *The Railroad and the Space Program*, E. Mazlish (ed.), Cambridge, MA: MIT Press, 1965.

[8] Fishlow, A. F., *American Railroads and the Transformation of the Antebellum Economy*, Cambridge, MA: Harvard University Press, 1965.

[9] Ringwalt, J. L., *Development of Transportation Systems in the United States*, Philadelphia, PA: Railway World Office, 1888.

[10] O'Brian, P. (ed)., *Railways and the Economic Development of West Europe, 1830–1914*, New York: St. Martin's Press, 1983.

[11] Fogel, R. J., *Railroads and American Economic Growth: Essays in Econometric History*, Baltimore, MD: Johns Hopkins University Press, 1956.

[12] David, Paul A. *Technical Choice, Innovation, and Economic Growth*, London: Cambridge University Press, 1975, p. 299.

16

Where Are We Today?

When beginning a journey, the direction one takes depends a great deal on where one is going. It also depends on where one is starting.

Before we tread our way into the future, we present a short sketch of where we are today.

We begin by summarizing the properties, capabilities, and soft spots in three of our intercity modes: road, rail, and air. Huge quantities of bulk commodities are also moved by water and pipeline, but we have little to say about these systems, because for most of us they are behind the scenes; we omit discussion of their characteristics.

The highway and road system

This is our only system that can almost always provide true door-to-door service for either passengers or freight. It's the only system that can move both small batches and moderately large batches, from pizzas to pianos to cargo containers, from Aunt Millie in her car to a busload of people.

One's personal car is on-demand service. There's no need to study schedules or wait at the station. It's ready when you are. For some people, it is their only chance for privacy in a very busy world.

Leaving aside walking and bicycles, it is our least expensive form of personal transportation because the driver usually performs for free. Even when the driver gets paid (taxis, school buses, airport vans, and such), it's still the cheapest form of public transportation for individuals and small groups (small batches in our parlance).

But debates about costs and usage color public policy and lead to shouting matches at public meetings. It is properly pointed out that automobiles do not pay their full costs. We have discussed congestion and pollution costs, and there is a large body of work on those costs as well as on the costs of accidents [1]. But there is still argument about what these full costs really are, and one looks in vain for analysis or even informed speculation about what would happen if all the modes paid their full costs. There is lots of partial analysis, but little that is comprehensive.

Everyone paid full costs? Full costs in rural Kansas is one thing and in downtown Chicago another. Even so, our guess is that things wouldn't change very much. There would be fewer buses in the suburbs and autos in congested downtowns. Intercity passenger train service would be greatly restricted.

What's wrong with road transportation? One, it's too slow for long trips. "Too slow" is relative; compared to the Conestoga wagon it is not too slow, but the airplane has radically changed our yardstick. Now many lifestyles are built around the availability of casual mobility over moderately long distances.

What do we mean by "casual mobility?" We have noted before that an hour is pretty close to most people's time budget for all the trips one makes in a day. Where uncongested freeway travel is available, that translates into maybe 50 or 60 miles for a day's supply of casual there-and-back trips. Now this "time budget" is a highly subjective and variable judgment that hinges on lots of factors, but an hour seems to be the point at which most of us start getting a touch reluctant to invest more time and hassle in routine travel. Not everybody, of course; masochism is not dead.

But there are lots of people beginning to wish the automobile was faster.

What else is wrong with road transport? It's too good, and it's too popular. This is a strange complaint to appear in the problem category, but popularity has a price. Part of the price is in the congestion it causes, part in the human and material cost of accidents, and part in the resources it requires.

The terrible state of urban traffic is really a different subject from our intercity discussion here, but it's not irrelevant; the total intercity trip has an urban component. Further, in sprawling urban areas the difference between an urban trip and an intercity trip begins to get a bit fuzzy, and we don't want to try to draw the line too finely. We've already addressed urban traffic congestion.

There will continue to be incremental improvements of one kind or another in the road-highway-vehicle system over time, but the existence of so many vehicles constrains how much the highway can be changed, and in turn the need to operate on the existing highways keep us from any drastic changes to the size or dynamic characteristics of the vehicles. And while these vehicles will very probably continue to become more fuel efficient and even less polluting, they will still depend primarily on petroleum for many, many years.

Trucks are unlikely to get much larger or heavier because highway design and safety concerns constrain them. Cars and trucks could easily get a little faster, but here both highway design and the capabilities of human drivers are constraints, although the intelligent vehicle kinds of things we have discussed will be very helpful. We've already discussed alternative car designs in Chapter 6, and we would not be surprised to see some of those possibilities actually come to pass for both cars and trucks.

The air system

Air is fast. It really is creating one world.

Except it is not very fast for short flights, because there is too much time eaten up on the ground.

Air is very, very rarely a door-to-door system. Nearly always a car or bus or urban rail or truck has to bring the payload to the aircraft and, at the other end of the trip, deliver it to the final destination. These are time-consuming actions that make the average, door-to-door trip speed markedly less than the speed of the aircraft itself. The penalty is much greater for short trips than long ones because the air portion is a smaller proportion of the total.

In addition to this pick-up and delivery time, only a part of the gate-to-gate time is flying at cruise speed. Taxi, takeoff, and climb out to cruise altitude, and then let down, landing, and taxi eat into full-speed cruise time. A 500-mph aircraft might take an hour to go 150 to 200 miles.

A 200-mile, door-to-door trip by air takes just about the same time it would take to drive.

But as distances get larger, air takes on a decisive trip-time advantage. The lost time does not increase any further, and the time spent in high-speed cruise grows accordingly. A 700-mile trip, a day and a half's driving for most people, might take half a day by air.

It is not an exaggeration to say the air service has created the mass market for long-distance passenger travel. And long-distance air freight, carried both in dedicated aircraft and in the underbellies of passenger aircraft, is already significant and growing rapidly for cargo that has a large time-value. In ton-miles carried, however, it is still minuscule compared to rail or road.

The squeeze point in air service is the airport, the interface with ground transportation. Airports are becoming almost miniature cities in themselves, with all the problems of small, dense cities: the need for people support systems, internal circulation, security, congestion. We talk about some of these in Chapter 22.

Air is indeed fast, faster than most of us could have dreamed 50 years ago. But it is fast only for longer trips; below somewhere around 200 to 300 miles it loses to the automobile or the truck.

The rail system

From its introduction in the 1830s and 1840s until road transport began to become significant in the 1920s, rail was effectively the only intercity transportation over land, except where canals or rivers permitted parallel service.

By being the "only" in every category, small batch and large batch, passenger and freight, to the next town or across the country, poor man and rich man, it was by default also the best in every category. Today the passenger car, the truck, and the air system have stripped it of nearly all those markets except the large batch movement of bulk commodities, like coal, concentrated copper ore, newspaper, grain, and some large manufactured products. The growing traffic in containers, however, is widening the heterogeneity of its market.

At its peak about 1920 there were over 250,000 miles of railroad track [2]. By the mid-1990s Class I rail mileage was down to 109,000 miles, shrunk by more than a factor of two and still declining [3]. In contrast, there are some 174,000 miles of federal interest highways, where most of the heavy trucks operate. There are another 690,000 miles of state highways and 2,200,000 miles of so-called rural roads [4]. There must be a million or so miles of local access and arterial roads in urban areas.

Rail is still the cheapest bulk carrier over land, roughly one-tenth the cost of carriage by truck on a ton-mile basis. Rail can offer door-to-door service in moving coal from mines to power plants and with large goods that move in large numbers—as long as the flow is large enough to justify track being put in place. In many other cases, however, something else, like truck or cargo ships, have to bring the payload—generally a shipping container—to the rail line, making rail only an element of an intermodal system.

Passenger rail requires very large passenger movements to be economically self-sustaining. The automobile, and more recently the air system, have co-opted most of these passengers in the United States, so there are only a few if any corridors where the train can still make marginal economic sense. (Outside the United States there are a number of very dense routes where service is profitable.) Compared to car and air, the number of people traveling by train in the United States is minuscule.

Rail is slow. Part of the reason is that heavy loads beat up the tracks so that high speeds aren't safe; part of the reason is that passing is restricted, so the slowest train often sets the pace; and part of the reason is that, except in very strong markets, frequency of service is usually low so that things waiting to get on board often have to wait a long time.

As was the case with the highway system, change in the rail system is also constrained by what is already there. One of the most fundamental decisions that has shaped the modern rail system was to put the rails 4 ft, 8½ in apart, a story we recounted in our history lesson in Chapter 14. As you may recall, this decision was made in England nearly 200 years ago, largely by George Stephenson as he began work on the Liverpool and Manchester railway.

Safety

We should add a word about safety.

Motor vehicle safety has improved markedly over the years. For example, fatalities associated with motor vehicles of all types has dropped from 5.1 per 100 million vehicle-miles in 1960 to 1.7 in 1995. Even so, the toll in deaths, injuries, and property damage is still large—in 1995 nearly 42,000 people lost their lives in motor vehicles crashes and accidents, some 95% of all fatalities from all forms of transportation.

The rate varies between urban and rural and with the type of road. The figures making up the 1.7 per 100 million total vehicle-miles range from 0.63 on urban freeways to over 3 on noninterstate rural highways and roads [5]. But even though the U.S. traffic safety record is among the best in the world on a per vehicle-mile basis, motor vehicle deaths are still roughly equivalent to a major airline crash every other day. There is still much room for improvement.

Air system fatality rates are much more erratic from year to year than is the case with motor vehicles just because fatal accidents can involve far more people, but there has clearly been an improvement over time. Numbers have ranged from 44 per 100 million aircraft-miles in 1960 and 15 in 1965 to an average of 1.2 for the 5 years ending in 1995, which is actually better than the 1.7 average for motor vehicles. But given that air trips are much longer than auto trips, the per trip comparison is much worse. Perhaps a more meaningful measure of air safety is the fatality rate per aircraft departure, which was about 1.9 deaths for every 100,000 departures of U.S. air carrier aircraft. While there are some 70% more fatal accidents per departure of commuter aircraft than with the major air carriers, the commuters are smaller and result in fewer deaths: 0.26 per commuter aircraft departure [6]. Anyway one looks at it, there is much need and room for improvement.

In 1995 there were about 85 fatalities—both on and off the trains—per 100,000 train miles. (Grade-crossing accidents are counted with motor vehicles and cause roughly 1% of those fatalities.) There were 26 derailments per 100,000 train miles [7].

About freight transportation

Flattering things can be said about freight transportation in general. For individuals it is no bother: it is out of sight and out of mind most of the time. But magnitudes calculated on a per capita basis may surprise you. Per person, there are about 14,000 ton-miles moved each year or about 40 ton-miles a day just for you! It costs you money, a couple of thousand dollars per year, but that is buried in the cost of goods and services you consume, so you don't think about it.

For that money, the individual has an enormous number of consumption choices available, as a visit to a grocery or department or any other large store will remind us.

Those who purchase freight transportation fuss about this and that, but by and large it is user friendly. Want to send a big heavy shipment to foreign country, a birthday present to Aunt Sally, a priority package across the nation, a tanker load of wine to France? No bother. A phone call to a broker or carrier or a visit to the post office will handle the matter.

Not only is the act of shipping easy, there are lots of choices of services. Unlike passenger transportation where the automobile dominates short trips and air long trips, for freight there is no dominant mode, just a lot of different markets with one or the other of systems best for each. On a ton-mile basis, railroads carry the most, followed closely by coastal and inland water services, pipeline, and truck. With their higher tariffs, trucks lead on an expenditures basis by a wide margin.

But there are problems

People looking from the darkside see trucks on the roads congesting traffic and banging into automobiles and pedestrians, long freight trains blocking intersections, lots of truck, rail, and air freight noise. There are capacity shortfalls here and there. Shipper concerns include theft, loss, damage, service quality, and the possibility of high tariffs where competition is limited or absent.

There are also some cost disasters for society. Lightly constructed rural and urban local roads suffer costly damage from heavy trucks, the St. Lawrence Seaway and much coastal dredging are expensive propositions requiring subsidy, and many of the Corps of Engineers inland water developments are also money sinks. There are more.

Dig a bit to a second level and we see lots of energy consumption and some very costly activities. What is the most expensive ton-mile movement? It could be the movement of diamonds and gold in armored cars. But for the individual it may well be moving 5 lbs of groceries in the trunk of your car from the store to home. That is about $200 per ton-mile, about 10,000 times more expensive than moving a ton of coal a mile on a railroad. Should that worry us? Probably not, but it should make us think.

It's not hopeless

These systems today are nothing like the systems of the past because there has been a continuous process of improvement. After all, the freight system responds to shipping needs, which respond to the market for goods determined by people making their own choices every day.

We see no reason that such improvements won't continue. People have already identified dozens of ways through which these systems are getting better, responding to both the new options improving technology brings and to new needs of our evolving society and economy.

But these systems are up against constraints that will foreclose some potential options for further improvement. The fact is that these systems—the auto-truck-highway system, the rail system, and even the air system—are mature in the sense that their further evolution is significantly constrained by the need to be compatible with what is already in place.

The basic constraints on each are the result of technical, institutional, and other decisions made or not made years and years ago. The auto-truck-highway system was shaped by concepts of size that are traceable back to the horse and wagon and to the internal combustion engine advances made in the decades just before World War I.

Do we really know that cars and trucks should be 6- to 8-ft or so wide operating in 10- to 12-ft lanes at sixty or seventy miles per hour? These numbers will change only slowly in the future because the existence of every part of the system constrains change in the other parts.

But being mature doesn't mean we've hit the end of the line, not all dimensions of improvement are foreclosed. And in spite of our not thinking about it, a good bit gets done even to our mature systems to provide a wider variety of services.

We have to remember that each of these modes has gotten to where we see it now through an almost continuous process of evolution. Most of the time change has been slow, seemingly minor increments that never made the front pages, but with a few very important jumps such as those that have taken us from horse power to internal-combustion power; steam locomotives to diesel; propellers to jets. We have moved from a world in which the telegraph was a great step forward into one in which we can't keep up with the rapidity of advance of new types of communication and control.

This evolution is far from over and, in fact, appears to be accelerating. We now go on to begin our deductions for the desired directions for both this evolution and for the very real possibility of entirely new modes.

Martian has been sitting there with his eyes glazed over. Stirring himself he remarks, *"In a blink of my eye I see all the things you have discussed and even more. What to make of the situation? To be sure there are some pimples and warts to be treated and demands are expressed politically for more investment. But one thing I have learned in my short visit is that just because something has a glorious history doesn't mean that more of it would be valuable. I've learned that we need to be concerned about the returns from incremental improvements. More doesn't necessarily mean better."*

Martian is correct, of course. Just putting technological fixes on problems and polishing what we have using new technology or expanding by building more track or airports isn't the prescription.

"There is more than that," interrupts the Venusian. *"You need to say less about modes and pay attention to the characteristics of service that folk want. Tell me what these are and how to improve them."*

Suitably challenged we will move on in the chapters to follow.

References

[1] Greene, D. L., D. W. Jones, and M. A. Delucia (eds.), *The Full Costs and Benefits of Transportation*, New York: Springer, 1997.

[2] *Railroad Facts, 1996*, Washington, DC: Association of American Railroads, 1996.

[3] Barger, H., *The Transportation Industries 1889–1946*, New York: National Bureau of Economic Research, Inc., 1951, p. 78.

[4] Bureau of Transportation Statistics, *Highway Statistics, Summary to 1995*, Washington, DC: U.S. Department of Transportation, Table HM-210, 1995.

[5] Bureau of Transportation Statistics, *National Transportation Statistics 1997*, Washington, DC: U.S. Department of Transportation, pp. 123 and 124.

[6] Bureau of Transportation Statistics, *National Transportation Statistics 1997*, Washington, DC: U.S. Department of Transportation, pp. 115 and 118.

[7] Bureau of Transportation Statistics, *National Transportation Statistics 1997*, Washington, DC: U.S. Department of Transportation, pp. 143–144.

17

Where Would We Like to Go?

A man's reach should exceed his grasp, but not by too much.

We have just reviewed our intercity systems as they stand today. Now we examine the issue of what new capabilities, or what improvements in current capabilities, might be the most fruitful directions for future evolution of the total system. In particular, if we were to develop some totally new systems, without any of the constraints we inherit from the past, what new capabilities would we like to see?

We know that we risk dangerous oversimplification. But maybe starting to paint with a large brush is not a bad thing: it should help in providing an initial focus. We will worry about qualifications and nuances if they seem important.

Based on both our vision of the future market and the state of the technology available to us, we suggest as our first goal the lowering of the cost of small batch movement. We think this makes sense in guiding the further evolution of existing systems as well as for entirely new ones. The rationale for this assertion is laid out.

Goal #1: lower the costs of small batch movements

Today every mode of transportation we know has rather strong economies of scale; that is, the cost of shipping something—whether it be a person, an automobile, a ton of coal, or a head of lettuce—decreases significantly as the size of the total shipment increases.

Recall the $200 per ton-mile that we imagined a person expending to haul a few groceries in the trunk of the car. A nutritionally rational masochist might say that one trip per year with the car stuffed with cabbage, beans, and unbleached flour would take care of nutrition requirements and vastly reduce ton-mile costs. Carry 100 lbs of groceries instead of 5 and the ton-mile costs drop to $10. Think of the money you could save.

This kind of dumb example only illustrates that it's almost always cheaper to transport things in large batches than in small, but the price paid is the inconvenience of only having infrequent "shipments" and in perhaps foregoing variety and choice. The later aspect is best illustrated by taking one step back in the logistic chain to the supermarket itself, which is always faced with the tradeoff among fewer items, larger total inventories requiring space, or more frequent shipments. We who tend toward the self-indulgent want variety and choice and we want it now, and we don't want transportation to be a serious limitation on achieving those desires; we don't want to be forced by markedly higher costs to accept major inconveniences.

The high cost of shipping in small quantities is the stick that drives shippers to try to transform small batches into large batches. The common technique is to let the small batches sit on the dock, so to speak, until a large batch has accumulated. If we want more frequent service, we have to ship or move in smaller batches. If flows are not extremely large as it is with coal, grain, and many other bulk commodities, we have to trade frequency of service for economies of scale—trade quality of service for costs.

So our goal is only partly concerned with just lowering the cost of the trip with our bag of groceries or the one with the computer we ordered the day before yesterday. The real payoff, we think, lies in the implications of that lowered trip cost: lowering the cost penalty of high frequency of service when flows are thin.

Why do we home in on this one? Because the world's businesses are undergoing a fundamental change in the way they operate, an essential element of which is the progressively increasing need for higher frequency of service transportation for thin flows. The whole transportation market is growing rapidly, but, consistent with this metamorphosis, small batch movements are one of the fastest growing segments.

An article in *The Economist* tells a very exciting story about how this revolution in our economy is coming about [1]. It will impact nearly every sort of economic enterprise, but of primary interest to us here are those enterprises that design, manufacture, distribute, and sell physical items—these are the enterprises in which transportation plays a major role.

In a nutshell, the advent of instantaneous and continuous communication through the Internet or Internet-type networks in combination with ubiquitous computing and so-called enterprise software is permitting a new kind of organizational structure along with changes in operating behavior and customer relationships. These new tools of information sharing and shaping are enabling multicompany business enterprises that are more tightly integrated, better coordinated, more cooperative, and more immediately responsive to customer needs and reactions than have existed in the past—indeed, that have ever been possible without these new tools of communication and computation.

The changes in operating behavior of these new organizational structures are sufficiently large in degree as to qualify as a change in kind: bringing into being an entirely new form of enterprise organization and operation, variously called "information-partnerships," "para-enterprises," or "virtual-corporations" in a "new business ecology." There are already visible examples of such highly productive and customer-oriented enterprises in this country, and all the portends suggest that this dramatic metamorphosis will spread worldwide.

Picture such a tightly integrated and information-cooperative enterprise manufacturing a line of products. The various elements of the business, the suppliers of raw materials and of components, the assembly facilities, the distribution channels, and customer interface elements can be anywhere on the globe. When a customer specifies an order, every element of the enterprise that is potentially affected instantly has the information to immediately react as is appropriate: components begin to flow, the product is created and delivered, and consumables are reordered. Customer reactions and practices feed back into the organization, resulting in immediate readjustments in the behavior of all the affected elements of the business.

It's very clear that this new economic environment is not one in which the various key entities can wait a week to let a big batch of parts accumulate on the shipping dock to save on shipping costs. Rather, one can picture the need for an almost continuous flow of individualized elements from one part of the enterprise to another and to their customers. It is a differentiated, flexible, just-in-time operation in spades.

It was interesting to us that The Economist's survey did not mention the role of transportation in this metamorphosis, yet it is clear that good transportation—both between enterprise elements and with its customers—is absolutely essential to effective functioning. Why is it not mentioned? We suspect because what we already have is pretty good and the possibility of anything significantly better is not imagined.

But, clearly, high-frequency transportation of thin flows of goods is the third leg, joining the Internet and the computer to make up the primary triumvirate of enabling technologies for this powerful wave of change. Lowering their costs can contribute to lower production costs and may in time enable still more efficient techniques and processes of production.

It's well recognized that time is money for high value goods. While these goods are being shipped or while they sit in inventory someplace—at the factory, on some shipping dock, in some warehouse, or on retailers' shelves—they represent investment that is not returning anything. And the more valuable the good, the more businesses are willing to pay to shorten the time lost in this logistics chain.

Thus, the implications of more tightly integrated, just-in-time operations put the spotlight on transportation as an integral part of the manufacturing and distribution processes. In this context any shortcomings go beyond just raising inventory costs but have the potential to more seriously impact total process costs and customer service. Conversely, if transportation can be significantly improved, the cheaper and faster it can become, the greater its contribution to improving total process productivity.

There is nothing new about these basic ideas. Using the word "communication" to stand for transportation, Jules DuPuit said in 1844 in the *Annales des Ponts et Chaussées* [2], "The ultimate aim of a means of communication must be to reduce not the costs of transport, but the costs of production." But he never dreamed how far the computer, the Internet, and better, faster transportation could take us.

(Incidentally, Jules DuPuit developed the concept of consumer surplus that is discussed in Economics 101. It describes gains to consumers that are not captured by producers.)

The recognition of these trends and their implications lead us to think that significantly lowering the cost of small batch movements might well be the single most important step we could take to increase transportation's impact on the nation's and the world's economic productivity.

Goal #2: significantly reduce trip time for both passengers and freight

What is the impact on travel behavior or on economic productivity of turning an 8-hr trip for goods delivery into a 4-hr trip? Or 4 hr into 2 hr? Would faster movements enable new forms of organizations, or patterns of commerce, or methods of manufacture that are distinctly superior to what is possible now?

When it took nearly a week for a man on horseback or a wagon train to go to a city 400 miles away, it might have seemed very attractive to cut it to a day. But people still underestimated the impact: "it can't make much difference, because very few people go there anyway." Circular reasoning, circling in the wrong direction.

Let's go back to our "why cut trip time from 8 hr to 4 hr or from 4 hr to 2 hr" and think of it a little differently. Today if one wants to make a trip or make a shipment that takes a day, the decision to do so generally has to be made the day before. If the trip time drops below a day, then the decision can be made the same day, as long as it's made before scheduled departure time and there is still space available. If trip time drops to 2 hr, a decision as late as mid-afternoon is good enough for a close-of-business delivery if a late-afternoon trip is scheduled.

This discussion implicitly assumed service was offered fairly frequently. If service is once a week, cutting time in transit from 8 hr to 4 hr is unimportant. Here the big improvement would come from increasing service frequency.

If service frequency is twice a day, then cutting time in transit from 8 hr to 4 hr could be very significant.

There is also the desirability of keying the times of service to fit the natural tempo of the process being served—this could be either people going to work or factories belching forth products. If, for example, goods are ready for shipment twice a day at 9 A.M. and 3 P.M., but the train or truck always leaves at 8 A.M. and 2 P.M., we've missed the boat. (Ah, the pitfalls of using metaphors!)

The other side of the coin is that if service frequency is really constrained—as, for example, flights across the Atlantic on the Concorde—some people will gear their schedules to match availability just to avoid the onerous of long flights.

There are obviously lots of unique circumstances. Clearly a little common sense is very helpful in finding the best way to optimize service.

This thinking leads us to believe that we asked the wrong question. A better measure of transport performance may derive from thinking in terms of the total time required to ship starting from the decision-to-ship rather than from the actual start of the shipment. Just thinking in terms of actual shipment time ignores the importance of frequency of service. Our first and second goals are now clearly part of the same animal.

We can be fairly sure that shorter trip times combined with high service frequency for people or faster delivery for goods will increase the demand for such services, but we don't know how much. When we push familiar things into levels of unfamiliar performance, we introduce great uncertainly in estimating the size of the market that will result. It gets us back to having to forecast the opening of market niches that we do not now foresee but on faith anticipate that they are there.

We repeat a point that we've made before and will probably make again and again in the future. Transportation is not an end in itself, it is an integral part of nearly all our economic and social processes. Therefore the implications of significant improvements go beyond the immediate effect of increasing revenue to transportation; the true worth of improvements come through the opportunities they provide for the introduction of new and better ways of performing other parts of these processes, the reoptimization of these processes. But in most cases these are far too complex and varied for us to be able to spot in advance the nature and extent of these improvements.

So how important is it to have transportation get so good that it fades from the decision process and stops being one of the drums that sets the tempo of commerce and manufacture? Our guess is that it is very important, but in ways that we cannot foresee without a more analytical imagination and a much more intimate knowledge of the myriad of processes that make up our economy. The more we lower the cost of small batch movement—and thereby increase service frequency—and speed actual shipment time, the closer we move to this happy state.

Sometimes necessity is the mother of invention, but if we invent ways to significantly lower the total time from decision to destination for high-value cargo—including people—invention may turn out to be the mother of necessity.

Goal #3: lower costs overall

We add this as an afterthought. We proudly offer the profound observation that, ceteris paribus, lower costs are better than higher costs.

But deep down, it is our feeling that if better performance is important, then people will be willing to pay reasonable premiums to obtain it. In the beginning, when costs of premium service are very high, only a few will pay. But if the new performance level becomes important for a few, then usage will spread to a few more. Over time, scale and experience will bring costs down.

There is wide band between "exorbitant" and "cheap." For goods with reasonable value per pound, we intuitively feel that if actual costs are "reasonable"—somewhere in the middle—then they are less important than service frequency, transit time, schedule reliability, safety, and a few other of the many dimensions of service. If the performance and properties are really desirable and useful, costs will take care of themselves.

But there are goods where cost is the dominant consideration. And there are a lot of them that travel in very large quantities. There are certainly real gains to be made by lowering their costs of transportation.

Other dimensions of performance

We have stressed cheaper and faster, but there are many aspects of performance and service other than those we have addressed: comfort for passengers, protection from damage for goods, safety, schedule reliability, and ability to handle a variety of shipment sizes, for example. We would never argue that any of these should be ignored.

All right, let's see what we can come up with in the next several chapters.

References

[1] "The Net Imperative," *The Economist*, June 26, 1999, p. 62.

[2] DuPuit, J., "On the Measurement of the Utility of Public Works" (translation of a 1844 paper by R. H. Barback), *International Economic Papers*, Vol. 2, pp. 63–110.

18

Serendipity

Stephenson and Pease must have engaged in a lot of "what if we do this or that" conversations. In the same spirit, we indulge in a similar exercise. We begin with a safety concern and see where that leads us.

There is one rough spot in the U.S. transportation system that we have only casually mentioned. We think it deserves a little more attention.

There's just no contest between a 2-ton car and a 40-ton truck. And trucks seem to be getting bigger as cars are getting smaller. In spite of the fact that most of the drivers of the big 18-wheeler trucks are usually both considerate and careful, a third of fatal accidents on highways involve trucks. And it's the automobile occupants who lose most often.

We have known for some time that the statements just made are in the ball park [1], and we also know that the situation isn't simple. Trucks come in many size and weight configurations, including, for example, suburban utility vehicles. We also know that drivers and driving environments are highly diverse. Cars and trucks mixing in light traffic on a high-quality freeway during good weather is one thing, but there are other situations. We know, too, that more than collisions are on folks' minds—noise, turbulence, blocking of sight distances, and passing difficulties are of concern.

Although any differences in weight and velocity aggravate crash outcomes, debates are mainly about more big trucks and big trucks getting bigger. Conclusions about big trucks get tricky when we remember that a large truck can take the place of several smaller ones. This might reduce the probability of truck and car conflicts while increasing the probability of large truck and car conflicts [2, 3]. There are big unknowns here.

As evidenced by letters to editors and news stories, there is a safety imperative and the general public just isn't interested in more and larger trucks mixing with its driving. That's the long and the short of it, the reality of it.

Another reality is that bigger trucks make a lot of economic sense. Double the size of a truck and costs decrease sharply. All other things being equal, which they rarely are, one driver hauls twice as much. Other operating and equipment costs decrease (fuel, trailer, and insurance) but less than by one-half. Shippers like these decreased prices, and there are potential reductions in pollution emissions.

These efficiencies have been studied over and over; and while exact numbers are debated, it is certain that incremental benefits greatly exceed costs as trailers are doubled and tripled and as weights are increased [4, 5]. The findings have been refined by economists who point out that appropriate tolls on heavy trucks would fund highways upgraded for larger trucks [6].

So there is a collision between the perceived safety imperative and the efficiency imperative as recognized by highway and bridge providers, shippers, and those interested in fuel efficiency and reduced pollution.

Separate highways?

What if we build completely separate highways for trucks and for automobile traffic?

This separation sounds like a good idea for both the truckers and the automobile drivers. It isn't a new idea. An editorial in 1928 stated that

"...nothing seems more certain than that many special highways will be constructed for motor trucking" [7]. Perhaps the time has come for that old idea.

Today, interstate highways are often being expanded anyway to cope with the growth in traffic in the heavy corridors, and this is where the problem is most acute. Indeed, separation got started in many urban areas as early as the 1920s and 1930s when truck routes bypassing downtowns were marked. Many city streets prohibit passage of even medium-weight trucks. Lots of parkways prohibit truck traffic. Why not just start building separate highways for the trucks everywhere?

Our unbiased and perceptive Martian visitor looks the situation over from his vantage point. He can easily see the highways, many of them with heavy streams of trucks mixed with the cars. And he can also see railroad lines connecting most of our cities—lots of rail lines, for the mileage of major rail routes is about twice that of the interstate.

But what catches his eye is whole trains carrying containers that look just like those on some of the trucks. (We call that COFC service, for container-on-flat-car). There are also those that come from container ships. He also sees some trains carrying the truck trailers themselves (this is TOFC service, for trailer-on-flat-car; we refer to either COFC or TOFC as "piggybacking"—a class of "intermodal" shipping).

The Martian makes what appears to be a very reasonable suggestion: *"You don't need to build expensive new highways just for trucks when you already have railroads that can carry the same things that most trucks are carrying or even carry the truck trailers themselves. Why not just let the rails carry the stuff between cities, and let trucks do the delivery of the goods to their final destinations once they reach the urban areas. This would get trucks out of the intercity business and eliminate the need for expensive new highways for intercity trucks."*

This sounds very logical, but our Martian gentleman just may be a bit quick in his prescription. If he looked back far enough he would see that this was the way it used to be. Trains moved intercity freight and horses and wagons and later trucks did the urban collection or distribution of freight. But as intercity roads were improved, intercity trucking grew. It has turned out that trucks have taken much of the intercity freight market away from the railroads, particularly for the higher value goods, and they continued to gain market share into the 1970s.

The fact is that trucks are able to offer much better service for such goods than trains largely because they are designed to carry the smaller loads that are more typical of such shipments and the road system gives much better universal access to origins and destinations than the rail network. We discuss this further later.

So the Martian is mostly seeing market share that has been saved from diversion to trucks or recaptured by the railroads as TOFC-COFC service.

But if he watched long enough he would also see that piggybacking trains operate on relatively few routes, routes between large-population areas where the flows of traffic are very heavy and where the distances involved are large, more than 600 miles or so; in fact, the average piggybacking movement is a good bit more than that [7, 8].

While intermodal movements have grown very rapidly in recent years, they still constitute less than 5% of truck ton-miles (although in 1997 they produced 17% of revenues, second only to coal at 22%) [9]. Smaller cities and towns are getting no very direct piggybacking service; they are being primarily served by trucks and sometimes by ordinary freight trains.

Why not TOFC or COFC everywhere?

The answer lies in the nature of trains. The basic problem is that trains are too big, strings of 50 to over 100 cars pulled by as many engines as necessary for that load and route.

But why is this "too big." Railroaders, like everybody, have to worry about lots of things—like safety, schedules, reliability, and how to protect cargo—but they are forced by competition to worry most about keeping costs low [10]. Having lots of cars per train is one of the primary ways of keeping the cost per car low.

And why does having lots of cars per train keep piggybacking from working everywhere? Our Martian could figure this out by himself by watching a shipment by COFC from origin to destination. (You can see a lot just by watching.)

The trip starts with the loading of the container on a truck trailer at the factory or warehouse where the shipment originates. A truck-tractor then attaches to the trailer and pulls it through the streets to the loading yard at the rail line. The container is taken off the truck, which then goes off to bring in other containers. The containers sit in the loading yard until at least a trainload of containers that share a common destination accumulate.

When a trainload of maybe 100 or more containers accumulates and an empty train with specially designed cars is available, the containers (or truck trailers in the case of TOFC) are then loaded. Many cars can carry multiple containers; the largest ones can carry as many as 10 containers, doubled stacked on five platforms. The full train is then ready to go.

How long does the first container delivered to the loading yard have to wait before it is shipped? It depends on the volume of freight from that point. If a few hundred containers per day per destination go through that point, then several trains per day can leave full.

But if it is not a large shipping point—perhaps only with the potential of a few containers per day—then it would take a week or more to accumulate enough to fill a train. Such a shipping point is highly likely to depend on trucks rather than trains.

Trains are just too big to serve small flows of freight.

Now picture the alternative: straight trucking. A container is loaded on a truck-trailer (chassis) at the factory or warehouse, attached to a truck-tractor, and the combination truck heads out down the interstate. It does not need to wait for other containers to be loaded or to be moved. In that driver's shift it can probably be driven directly to the unloading dock at its destination some 300 or 400 or even more miles away. (The average truck trip is a little over 400 miles; the average freight car trip about twice that [11].)

If our truck is going farther, then either a second driver takes over at the end of the first shift or the truck stops overnight and same driver resumes the journey the next day.

Truck-rail-truck intermodal shipments, with the time disadvantage it already has with the loading and unloading operation, can't even come close to competing on overall speed of delivery with direct trucking unless the distances get beyond what a truck can do in a day.

But at longer distances the contest changes. Now the trucks require two or three work shifts for their double-teamed drivers, and here the piggybacking can be competitive on service and its costs are much lower.

But even at these longer distances, piggybacking can't compete unless the flow of traffic is heavy enough to support reasonably frequent train service. These conditions do obtain lots of places, and where they do piggybacking is the prevailing mode; the Santa Fe estimates that some 90% of the long-distance truck trips between California and the midwest go intermodal by rail [12].

Railroading is at its most efficient when there are enough cars going from Point A to Point B so that they can be made up into one train and the whole train can make the trip. Shipments of individual cars from Point A to Point B are also handled by the railroads, but less efficiently and slower since each car or block of cars must usually be transferred from one train to another to finally get where it is going. Taking trains apart and putting them back together to send each car in its desired direction is both time-consuming and expensive.

Trains are very good at moving bulk commodities like coal or grain because they move in large volumes and permit the operation of these "unit" or "through" trains.

"Why not put all the trucks on trains indeed! Isn't that just like a Martian, shooting from the hip like that? Not only won't piggybacking get very many of the trucks off the highway, trucks are absolutely necessary because rail provides such lousy service for small freight flows and shorter distances. Many smaller towns get no train service at all," the Venusian mutters in an unusually long mutter.

Our now chastened Martian doesn't give up easily. *"What if you made the trains smaller, with fewer cars. Then you wouldn't need such large flows to make piggybacking efficient. Why not make just one-car 'trains.' Or maybe just a few cars per train. What if you develop some kind of a minitrain that only carries a few containers or trailers, and you develop transfer facilities tailored to these minitrains that are faster and cheaper? With smaller trains the container or truck trailer wouldn't have to wait so long for a whole trainload to accumulate. Even though these short trains will probably cost more per ton-mile than a 100-car train, they might still be a lot cheaper than building a whole new interstate system just for trucks."*

Now he may be on to something. How did we get this way, with trains so large that we can't really give good service for "small" shipments? The primary reason is that we're still in the same technical format that was established when railroads came into being some 175 years ago.

As we described earlier, the early steam locomotives were nothing but substitutions for the men or animals that were pulling coal carts and wagons from mines to the canals in England. It turned out that these first locomotives were powerful enough to pull several "cars" at once. It didn't take long to start coupling the wagons together, and "trains" were born. The scale economies of lots of cars per locomotive and its crew have kept them that way.

Further, this was really the only format that made technical sense. Steam engines were big and inefficient, used bulky fuel, and took a couple of men to operate. They just took up too much space to even think about putting a steam engine on each of the cars that carried the payload; it only made sense to put this bulky and complex power source and its operating crew on its own platform, and the freight or passengers on other vehicles strung out behind. Remember, too, that the first locomotives were built to pull coal carts and wagons; the idea of putting an engine on each of them rather than on a separate platform made no sense whatsoever. And practical internal combustion engines that could be made more space-efficient were still over 50 years in the future.

There was also the problem of controlling train movement, keeping trains from running into each other. Putting lots of cars on the train eased control problems.

As steam engines got better and could produce more power and pull more cars, the obvious thing to do was to make the train longer. For about the first 100 years of railroading, the pattern of evolution was progressively larger and more powerful locomotives pulling greater and greater payloads. In recent decades the increase has been through larger cars; the number of cars per train has actually been decreasing very slightly.

Now the Martian suggests we play another "what if" game: what if we sacrifice some of these economies of scale and make trains a lot smaller rather than larger.

Minitrains? Individual, self-powered cars?

Yogi Berra is credited with giving the dubious advice that when you come to a fork in the road, take it. We hit such a fork back in the 1930s, but it is highly doubtful that anyone even noticed it, let alone thought about it. This is when we began replacing steam locomotives with diesel power. The diesel has not only produced a lot of nostalgia for the good old days of huffing and puffing railroading, it has also produced a very big improvement in operating costs. Today nearly all trains are powered by diesel internal combustion engines (actually, diesel-electric—diesel engine driving an electrical generator; electric motors driving axles).

So where was the fork in the road? One was the option we took to substitute diesel locomotives for steam locomotives. Trains kept on being trains and kept on getting bigger.

The fork not taken, or perhaps not even recognized, was to do what our Martian friend suggested: to move toward smaller trains, even to single cars that carried their own power. This opportunity stemmed from the fact that it was easy to scale diesel engines to build them to the size needed to power smaller trains or just individual cars. Along that fork "trains" didn't have to keep on being bigger, longer, more powerful; they could get smaller and more flexible—like trucks, only on rails.

There are at least four reasons we didn't take that fork back then. First, railroaders didn't think that way. Although truck competition was real and growing at the time, in the late-1930s many managers didn't really recognize the competition that was to come from the still relatively new trucking industry. Railroaders did not see that trucks, with their ability to carry small batches and offer faster service, would steal the high-value goods market, leaving them with only the large batch and bulk commodity market.

Second, the railroaders were quite properly concerned with costs, and smaller trains brought higher costs. The labor unions left no way at the time to avoid fairly large crews, and large trains helped absorb their costs. Small trains would have been totally inconsistent with low cost. (Even in 1995 cost per ton-mile for trains was roughly one-tenth that for intercity trucks, about 2.5 cents for rail versus 25 cents per ton-mile for truck.)

Third, the railroads' traffic control problem would have been a nightmare. It was tough enough making sure that the relatively few trains sharing the same sections of traffic didn't run into each other; the problem would compound with an order of magnitude with more mini-trains or single, self-powered cars running around on the same tracks.

Last, when the substitution of diesel for steam was really ready to get underway, there was a war of some size going on. The 1940s were not the time for railroaders to be distracted because of concern about the future of their market. Indeed, wartime controls prompted the building of a new round of modern-for-the-times steam locomotives, which may have delayed the introduction of diesels. After the 1940s dieselization took off rapidly, and in fairness to the railroad community, there were leaders who saw potentials for marked service improvements [13].

"Just like a Martian to oversimplify the problem," the Venusian thought to herself.

Her judgment may be too quick, too. That was then, and now is now, and maybe today things are different. Attitudes of railroaders are certainly very different.

Second, automation can cut down on crew sizes, and it is becoming feasible to think in terms of completely automated systems with no crew on board; in fact, there are fully automated trains today carrying coal at the 4-Corners power plant. Some trains in Canada are operated with one-person crews. Eliminating, or even just reducing crew size, could help reduce the cost penalty that comes with smaller trains.

The traffic control problem is well within our reach to solve, given all the new sensing and communications gear and systems like the Global Positioning System that can determine a train's—or a just rail car's—position within a few feet.

We can think of no fundamental technical problem standing in the way of the Martian's idea.

What does that kind of single vehicle or minitrain capability gain us?

For one thing, it makes it possible for the railroads to offer reasonably good service with smaller batch movements—either people or freight. It means we can have more frequent service and thus reduce the time that any movement—people or freight—has to wait for carriage. If, just to illustrate, the volume of traffic will support one 60-car train every 24 hr, then it will support an individual car every 24 min and, once loaded, that car can go directly to its destination.

If train sizes are more flexible, then the "train" can be tailored to the size of the shipment or passenger load and the frequency of service desired. Because it can give good service with smaller shipments, there is an opportunity to recapture some of that higher revenue market from intercity trucks.

Obviously a lot of this hinges on how the relative costs and service turn out.

Unless someone can think up something novel, minitrains would still probably require that the shipment be already on the rail line or be brought there by truck, so the truck still wins on point-to-point access.

One such development is worth mentioning. Truck-trailers have been devised that can be driven onto a rail track and have rail wheels inserted under them. Subsequently, they are pulled in trains as if they are rail cars. They may be demounted and pulled by a truck-tractor to the final destination. So far, this service has appeared in market niches, as we would expect. There is still much room for innovation in the 170-year-old rail mode, and there is much innovative activity.

The Martian was proud of his idea and pleased by our favorable reaction. What he still has to learn is that it's very hard to have a new idea that somebody else hasn't already thought of. In fact, both the Danes and the Germans have already moved in the direction of self-powered cars—and maybe others have by now.

The Danes have built a passenger train called the Flexliner with cars that can be operated individually or coupled together to form trains. And this coupling or decoupling can take place while the train is moving. Thus the Flexliner cars can form up into trains en route, or come apart as need be. And the Germans have three-car minitrains for freight movements in initial service.

A drive-on, drive-off train has been proposed and tested in the United States and Canada—an iron highway concept. It is imagined that it would serve in markets where traffic quantities are limited and/or haul distances are short. Either dock-to-dock or collector-distributor services to TOFC-COFC hubs might be provided. In one version, the sides of cars fold out to make ramps so that loading and unloading can be done anyplace there is room to maneuver vehicles.

Several U.S. and Canadian railroads have offered services using diesel-powered single- or multiple-car rail-bus formats (Budd cars), but they gave way as markets eroded. Yesterday's interurban streetcar-like systems had a similar format. Automation and control is what is different today.

Back to our original problem—getting the trucks off the highway

We defined for ourselves two basic options to make highway travel easier and safer for cars and probably more efficient for trucks.

The first is to build new highways exclusively for trucks. There is no need to dwell now on all the pros and cons of this move.

The second is to expand the rail system and introduce new kinds of small trains along with a new system for operational traffic control. But, as we will explain, while it may make rail more competitive with rail, we doubt very much that it will have a sufficiently significant impact on the number of trucks on the road to solve the safety problem with which we started.

The minitrain and the self-powered single vehicle would appear to make the rail system more competitive with trucks for thin freight markets and could reasonably be expected to capture some portion of that market. It also distinctly broadens the conditions under which piggy-backing is preferred to straight trucking, both by making it feasible in thin markets and by shortening the distance at which it might become attractive. With only long trains, rail was almost completely noncompetitive with the truck in thin markets.

While the minitrain or the individual self-powered car can provide much better intermodal service than today's long trains, it still can't beat trucking on total trip time under about 300 to 400 miles. The average truck haul today is a little over 400 miles, so for roughly half of the truck trips piggybacking is still at a service disadvantage, but a much smaller one.

A big unknown is relative costs. We won't even try to guess.

No matter how they turn out, however, the hill we are trying to climb to significantly reduce the number of trucks on the highway is a very steep one, given that intermodal shipments make up such a small portion of truck movements today. Even if we double or triple the intermodal market we have only decreased highway ton-miles by something like 10%.

So it appears that unless our new rail service is successful beyond our wildest dreams, we will still be left with too many trucks sharing the highways with cars. We are left with the tentative conclusion that if we want to avoid the troublesome, even dangerous, mix of large trucks and smaller cars on the same highways we might be forced to build separate highways or find some other way to physically segregate the traffic. Maybe it might not have to be done on all highways, but certainly on a lot of them where the combined traffic is heavy.

But wait!

Except for a mumble now and then, our Venusian has been quiet through all this. She now speaks up, *"Before we start construction on the new highway system just for trucks, my intuition tells me that we ought to think about this a bit more."*

Always accommodating, we agree to do so.

References

[1] Eicher, J. P., et al., *Large Truck Accident Causation*, Report 806-300, Washington, DC: U.S. Department of Transportation, National Highway Safety Administration, 1982.

[2] Transportation Research Board, *Twin Trailer Trucks: Effects on Highways and Highway Safety*, Special Report 211, Washington, DC: National Research Council, 1986.

[3] *The U.S. Department of Transportation's Comprehensive Truck Size and Weight Study, Volume III DRAFT*, Washington, DC: U.S. Department of Transportation, 1998, pp. 8-4 and 8-5.

[4] Winfrey, R., *Economics of the Maximum Limits of Motor Vehicle Dimensions and Weights, Vols. I and II*, Washington, DC: Federal Highway Administration, U.S. Department of Transportation, 1978.

[5] Solomon, D., et al., *Summary and Assessments of the Size and Weight Report*, Washington, DC: Federal Highway Administration, U.S. Department of Transportation, 1981.

[6] Small, K. A., C. Winston, and C. A. Evans, *Road Work: A New Highway Pricing and Investment Policy*, Washington, DC: Brookings Institution Press, 1989.

[7] McKenzie, D. R., M. North, and D. Smith, *Intermodal Transportation: The Whole Story*, Omaha, NE: Simmons-Boardman, 1989.

[8] Federal Railroad Administration and Maritime Administration, *Double Stack Container Systems: Implications for U.S. Railroads and Ports*, Washington, DC: U.S. Department of Transportation, 1990.

[9] Association of American Railroads, *Statistics and Facts*, Washington, DC: Annual. (Accessible online at http://www.aar.org/comm/statfact.nsf.)

[10] Armstrong, J. H., *The Railroad, What It Is, What It Does: the Introduction to Railroading*, 3rd ed., Omaha, NE: Simmons-Boardman, 1990.

[11] Bureau of Transportation Statistics, *National Transportation Statistics 1997*, Washington DC: U.S. Department of Transportation, 1996, Table 1-10.

[12] Bureau of Transportation Statistics, *Transportation Statistics Annual Report 1995*, Washington DC: U.S. Department of Transportation, 1994, p 25.

[13] Barriger, J. W., *Super-railroads for a Dynamic American Economy*, New York: Simmons-Boardman Publishing Company, 1956.

19

Beyond Railroading: a New System

"Genuinely new technologies are upon us."

Drucker, *The Age of Discontinuity*

The last chapter left us with the tentative conclusion that if we want to avoid the troublesome, even dangerous, mix of large trucks and smaller cars on the same highways we might be forced to build separate highways or find some other way to physically segregate the traffic. Maybe it might not have to be done on all highways, but certainly on a lot of them.

The Venusian has offered a cautionary thought: *"We are sitting here talking about maybe spending billions on a lot of new highways for trucks. The Martian's idea of making trains smaller—even down to individual cars—is really the only alternative we've thought of. That might be a pretty good idea, but we don't think that it can solve this big truck–small car safety problem. Before you make this huge investment, shouldn't we see if we can't think of something better?*

"Let's act like we have a perfectly blank slate and see what new characteristics and capabilities would best complement the auto-truck-highway, the rail, and the air systems that we already have. And getting trucks out of the automobile traffic stream should not be the only consideration. There should be lots of possibilities for improvement."

She stopped, a little embarrassed to have said so much.

But she, of course, is right. We are inevitably facing very large new investments in the maintenance and expansion of our existing systems. It certainly makes sense to look at how our existing systems might be improved but also to explore the possibility of doing something entirely new. It is only sensible to take advantage of all the new technology that time has put in our lap as well as the knowledge gained from our experience with the existing systems. We may well decide that improving the old is still a better bet than introducing something entirely new, but we will have made that decision in full view of all our options.

The key thing, we think, is to not continue down the same old paths out of blind inertia.

The Venusian suggested that we *"act like we have a perfectly blank slate."* To the technologist a blank slate is almost irresistible; it's easy to get carried away. In nothing flat we could be working on coast-to-coast pneumatic tubes or all kinds of things just short of "Beam me up, Scotty." We will resist the temptation and exclude from our scope technology that is still in the lap of the gods or schemes that we judge have little chance of ever coming into being.

The fact is that no slate is really blank, in the sense that there are still all kinds of constraints on what we do: technological, economic, environmental, public acceptability, and probably others. And it would be foolish to just arbitrarily reject what's in place now: much of it we might see no real reason to change, and it makes sense to build on what we can. Actually the line between "totally new system" and "major evolutionary change" is very vague, and we do not want to constrain our thinking by trying to define it.

We will, however, restrict our purview to systems that have some chance of growing into a nationwide intercity network. As we will discuss, it's unrealistic to think of building such a network all at once—they grow over time. Remember, the interstate highway system is over 50 years old and we are still expanding it; from inception to maximum rail track in this country was about 90 years. We would probably never have started either if the total bill had been presented at its beginning. But we don't want to include approaches that are only good for filling limited niches without some prospect of expansion.

One possibility we have in mind is a new "superspeed" highway, which we will examine in the next chapter. In this chapter we will focus on another class of options commonly referred to as fixed guideway systems.

We've never liked the appellation "fixed" guideway; all guideways are "fixed" in the location sense—none of them can get up and walk around. The phrase only means that the directional control of the vehicle is provided by the guideway, like on a railroad. Even though it's fairly immobile, a highway doesn't qualify because the vehicle supplies its own directional control; there is no "guideway" that steers it.

The Martian's advanced rail system—with train sizes tailored to the load—that we described in the last chapter is a fixed guideway system. While it would represent a major change in railroading, we would still somewhat arbitrarily view it as an evolutionary change, not a totally new system. It continues to use conventional rail, so vehicle widths and heights are largely limited by the familiar 4-ft 8½-in gauge and the clearances afforded by existing tunnels, bridges, and other constraining structures. A new system could shed these constraints and offer the opportunity for guideway designs that are entirely new.

There are a number of possibilities. A very interesting option is the exploitation of the magnetic levitated technology that has been in development for over 20 years, primarily in Japan and Germany, although much more recently there is an American system in early development [1–3]. There are advocates of these maglev systems who propose speeds over 300 mph. Prototypes of such systems are operating today and being planned for a few niche uses early in the twenty-first century. Interest in the United States has waxed and waned over the last 30 years [4].

Thinking about fixed guideway options

We are all familiar with highways. We use them all the time: they are part of life for most of us. While not everyone would agree that making them faster is desirable, very few would fail to understand why others might like the idea. The point is that we know how to think about highways, or at least think we do.

That situation doesn't prevail when we start talking about 200-mph trains or 300-mph maglev systems. The first thought of most people is that "when people start using this wonderful new system there will be fewer cars on the highway to interfere with mine." (I need mine because where I'm going is a long way from the station and I don't like taxis.) Most of us would hardly think of the implications of higher speed for freight movements.

Instinctive reactions are unlikely to be valid assessments of the potential of these new possibilities. For most of us this is unfamiliar ground we are plowing. We are envisioning new systems that will operate in performance regimes well outside of any previous experience. The egocentric view we might take of highways is inadequate to evaluate the potential of an unfamiliar type of system as a commercial venture or as a tool to improve our world.

So before we get into a discussion of specific designs, we are going to start with motives: why the world might want such systems, what market niches might they occupy, and what system characteristics are implied to fulfill these wants?

We hark back to Chapter 17 where we identified the directions of performance evolution that, in our judgment, we thought would be most useful in the future. We said that our first desire was to lower the cost of small batch movements and our second was to speed delivery. We'll start there, in the context of how these desires might influence actual system choices.

The implications of our desires

One implication of this desire is the need for our new system to support vehicles—whatever they turn out to look like—operating singly or in flexible combinations, vehicles that are or can be tailored to carrying small quantities of goods or small numbers of passengers. We won't try to define what these vehicles look like or what they move on, nor what we mean by "small," just yet.

To keep costs down the system needs to be highly automated. So we see this as central to any new system we define. We would like to automate as many aspects of operations as we know how. Add to that efficient vehicle design, flexibility of operations, and, of course, safety and environmental acceptability and we've just about summed up the main properties for which we will strive.

Lowering the cost of vehicle and system operations is only part of the problem. We still have to build the infrastructure—the track or guideway, the operational control system, the maintenance and support facilities, the stations and parking lots, and the offices, for example.

And, unfortunately, we have to pay for them. The only source of revenue is customers paying for vehicle-miles. So we've got to add to each vehicle's operating cost its share of these infrastructure costs. We do the same with the cost of the organizational overhead: management, personnel (now more popularly called "human resources"), accounting, sales, and stockholder relations, for example.

For a new system, it's unlikely that these costs will be small. This means that the only way to keep them reasonable on a per vehicle-mile basis is to have a lot of vehicle-miles. So we should from the beginning aim for systems that can serve as many different markets as we can identify: big batches—not coal cars, but maybe container size shipments—as well as small batches, perhaps as low as 20 to 40 people at a time, similarly sized vehicles for freight, with many classes of service. Big infrastructure investments can't be covered with thin flows. We must attract large flows made up of many, many small batch movements and very possibly some of the large batch flows railroading supports today.

So to lower the cost of small batch movement we need: (1) vehicles designed for small batches; (2) low operating costs, achieved through general efficiency and the liberal use of automation; and (3) a healthy market.

The Martian, as he is often wont to do, has something to say: *"Look, way back I proposed my improvement on today's railroad system. In fact, you mentioned it at the beginning of this chapter. It seems to me that it does everything you have asked for. It was specifically configured to be able to carry small batches, and it was going to be highly automated. And it doesn't need to attract a huge market because it shares the infrastructure with the large trains on the existing rail system. Why do you keep talking about a new system?"*

We have to admit that there are only a few really new features we can offer in a different system. We can move to a different kind of track, reducing the risk of "derailment" to almost zero, so we can operate safely at much higher speeds, and we can free ourselves from some of the geometric constraints of rail.

We've also made the case—well, we've at least identified the case— for significantly reducing the decision-to-destination time for both people and freight. Based on the past reaction of markets to faster delivery performance, we've conjectured that still higher speed would further expand the market, perhaps enough to justify an entirely new system. But we really don't know; we will discuss this point a bit more.

We discussed in Chapter 17 why we consider trip and delivery times to be important and brought up there the notion that in some cases the clock should start with the decision to ship rather than when the shipment actually starts: we focus on decision-to-destination.

We asked ourselves the question: "How important is it to have transportation get so good that it fades from the decision process and stops being one of the drums that sets the tempo of commerce and manufacture?"

And answered: "Our guess is that it is very important, but in ways that we cannot foresee without a more analytical imagination and a much more intimate knowledge of the myriad of processes that make up our economy."

While we can't quite get that good, we think that moving in that direction is the only way to attract a market large enough to justify a new system. Further, as we noted, we think that the market will be dominantly high-value freight, with people perhaps important but secondary.

The three ingredients that determine decision-to-destination time are frequency of service, ground handling and nonproductive trip time, and "cruise" speed. We will look at each.

Frequency of service determines the lost time the shipment spends sitting waiting for service to be provided (or the passenger in the waiting room). We repeat once more that high-frequency service is enabled by low costs for small batch movements. This, as we have noted, was the point of the Martian's advanced rail variant of today's railroading. Flexibility and a better small batch capability is at the heart of trucking's competitive advantage over rail; the Martian wanted to enhance rail's ability to compete with these same properties.

The second element is the time required to pick up the shipment and, if necessary, deliver it to our new system; and then, after the line-haul trip is over, the time to deliver to its destination. We must clearly do all we can to facilitate this pick-up and delivery, but it's hard to see why a new system might do better with this aspect than the Martian's system.

(This fixed guideway approach will require highway vehicles to perform the pick-up and delivery function simply because the high speed intercity vehicles designed for it are very unlikely to also be able to operate on the roads and highways in the cities. This is a big disadvantage to this kind of system compared to a highway-based system.)

Only the third element of total decision-to-destination time—the line-haul transit time—depends on the speed of our new system.

There is more to speed than just speed

Aside from the obvious point that speed shortens trip time, there are other considerations. In fact, it is these other factors that may well play the more important roles in the final choice of a number.

Improving productivity means getting more output per unit of input. Our "unit of input" is a vehicle of a given size; its output is miles per day. Because a fast vehicle can cover more miles per day than a slow one, going faster buys greater productivity of the vehicle and its crew (if any). The cost of air travel is low not because airplanes are cheap, but because aircrafts are very fast and "produce" a lot of seat-miles per day (and because people are too often mistaken for sardines).

Of course, speed is not the whole story. Ground systems don't have to go as fast as airplanes to generate the same miles per day (or to be competitive on delivery times). Ground system can beat aircraft in total trip time out to substantial distances just because the time required to load, accelerate to speed, decelerate to stop, and unload are very short compared to the analogous actions with the air system.

On the negative side, faster vehicles use more energy and generally cost more to build than slow ones. So we have to balance these increased costs against the advantages of increased productivity.

So optimizing a "let's start from scratch" system we expect to be operating for the next 100 or so years (remember: the railroads are over 170 years old now) is no small challenge. There are too many uncertainties. What will be the price of petroleum 20 years from now? Will we use it? How much larger a market can we capture with 200 mph instead of 100 mph, or 300 mph instead of 200 mph? How much will the general economy gain with these various levels of service? Even a very brief divine revelation would really help sometimes.

While we do look at alternatives in later chapters, we have concluded here that the primary dimension of improved performance we would achieve with a new intercity system is higher speed. We are also no longer restricted to two steel rails 4 ft, 8-½ in apart: we can change the guideway geometry. We can select designs that are not susceptible to derailments, and new "gauges" might open up now-unattainable options in vehicle design. All the other forms of improvement we have thought of can probably be retrofitted into our existing systems as a normal part of their evolution over time.

We sense that these changes are important enough to deserve serious consideration but admit that proving their ultimate worth with hard numbers is beyond us: our crystal ball is just not that good. But accepting that view as a given based on intuition and faith, we go ahead to examine some of the broad design considerations in choosing such a system. We are not trying to pick a design, just to illustrate the broad nature of our options and illuminate some of the considerations that will go into a final choice.

Networks versus closed circuits: the need for switching

From time to time one sees a picture of some streamlined-looking vehicle riding on a single rail like at Disney World, or hanging from an overhead rail, or riding in a U-shaped concrete track like a bobsled, or just some kind of new train riding on the familiar pair of steel rails. They all look very modern and impressive.

But it pays to look for the small print. What is seldom shown (except for rail) is any kind of switching system that would permit the vehicles to do anything other than run around in a big circle or just go like mad for miles and miles along a straight track.

This switching capability is critically important for any system that is going outside the amusement park or the airport. Without switching from one guideway to another we can't have a network, a vehicle or train can't stop for loading and unloading without forcing following vehicles—at least close ones—to stop also. Without the capability for one vehicle to pass another we force all traffic down to the speed of the slowest vehicle. We can't think of any way to do these things unless we can easily and quickly switch from one guideway to another.

So if the system is expected to do more than just connect a few points, we will need the ability to do reasonably high-speed switching in whatever new fixed guideway system we might develop. There are obviously other properties we will demand: it must be safe and relatively easy to build and maintain, for example. But this ability to switch must almost always be on the list.

Guideway options

As we noted, our new guideway doesn't have to look like rail. In fact, since we plan to operate at very high speeds, we would like to have a guideway that makes accidental "derailment" much more difficult than with steel rail as we know it. There are high-speed trains operating in the world today at high speeds, but they require almost heroic maintenance to remain safe. We would prefer a scheme less vulnerable to the deranged mind, or just the unthinking prankster.

We think there are two properties that would improve derailment safety. The first is a guideway design in which the vehicle is physically constrained to stay on the guideway even if something goes wrong so that if a dislocation or "derailment" of some kind does occur the vehicles don't go flying off into the countryside. The second—certainly not foolproof but probably helpful against the casual mischief maker—is elevated track. Now anything can be elevated, but lightweight systems would make the job easier and less costly.

One can envision a wide variety of alternate guideway design schemes. One is a channel with the sides high enough to prevent any possibility of the vehicles getting outside the channel. We won't try to decide whether the vehicles are riding on steel wheels, rubber tires, a magnetic field, or a cushion of air; that's a decision a bit downstream of where we are now.

Another option is a T-shaped guideway. The vehicle rides on top of the T, but with parts of the vehicles wrapping down around the edges of the T, physically constraining the vehicles from leaving it. This is the approach used with the so-called attractive maglev propulsion system developed largely in Germany; the guideway magnets are under the edge of the T, and the vehicle magnets are on the part that wraps around the T.

(Since magnets on the track are on the underside of the T and the vehicle magnets are on arms that wrap around it, the vehicle magnets are below the track magnets. Thus they attract to hold the vehicle up—if they repelled, it would pull the vehicle down, not hold it up. By sheer coincidence, this approach is described as an attractive system. The Japanese system, on the other hand, has the vehicle riding inside a channel lined with magnets. These magnets repel the magnets on the vehicle, thus holding it up. [Out of politeness, however, we don't refer to it as a "repellent system"]. The American system also uses a magnet in the vehicle that repels coils in a box beam guideway around which the vehicle folds. The guideway configuration changes where switching takes place.)

Another approach is an upside down, lower-case t. The vertical segment protrudes into a deep slot in the vehicles, which constrains the lateral movement of the vehicle. The wheels or whatever ride on the horizontal segments of the inverted T. Some experimental air cushion vehicles have used this approach. (Air cushion seems to have lost out to magnetic levitation.)

If we decide that very wide vehicles are desirable, like 10 to 20 ft, we may need dual "rails" or an equally wide base for the T or the bottom of the channel.

We might note here that the channel guideway with vertical sides appears to be the easiest to "switch" because if the vehicles are slightly steerable they can determine which branch to take merely by guiding on one wall or the other: guiding on the left wall leads into the left channel and guiding on the right wall leads into the right channel.

The other schemes, insofar as we can see, require physical movement of the guideway, much as tracks are switched today. Here is one constraint on the very wide system: as the guideway gets wider and more massive, the problem of switching is made more difficult. Unless, of course, somebody can come up with a brighter idea than any we have had so far.

There is one other approach to guideway design in which we shift the directional control of the vehicle from physical constraint by the guideway to lateral steering by the vehicle itself. We envision the use of electronic guidance to constrain the direction of motion: electromagnetic signals of one kind or another are used to mark the desired direction of movement and the vehicle steers to follow. Our guideway can now be just a flat surface on which the various vehicles are guided by electronic signals, not physical constraints.

The Venusian politely interjects: *"I apologize for this interruption, but it appears that you have just reinvented the highway, only adding a bit of automation."*

Caught again. We were really trying to sneak in another idea here: the possibility of a kind of hybrid guideway, in which the lateral constraint is physical for the very high speed operation, but electronic for switching and docking. The notion is to have the vehicle, perhaps after slowing, leave the physical constraint into a section where switching is done electronically. That section may be just a branch point so that as soon as switching or loading and unloading has been completed, the vehicle resumes operation on the portion of the system where the directional constraint is physical.

Maglev

We admit to a bias toward maglev, though admittedly without the depth of current analysis the subject deserves. It seems to us that in pushing rail to higher and higher speeds we are pushing rail technology to the end of its rope. Maglev is still a technological teenager, with its true potential just coming into focus.

The advantages of very high speeds are quickly compromised if much time has to be spent in stopping. So the natural market is where there are long straight shots, like shuttles to airports.

At 300 mph a large new international airport 100 miles out of town is now just over a 20-min ride away. It seems very possible that more and more airports will be 50 or 100 miles out of town in the longer range future. We can begin to see more possibilities of one airport serving several cities. We discuss this subject more fully in the air transportation chapter that follows.

How such initial niche installations expand in the future to form networks seems to us to hinge at least partly on the ease with which switching can be accomplished, a now familiar subject to the reader.

Very high speed: a fly in the ointment?

There are other issues that go beyond cost when we contemplate new higher speed systems.

Because the aerodynamic drag goes up rapidly with speed, so does the energy required per mile. As a rule of thumb, doubling speed increases the aerodynamic drag by a factor of four. Even though further stream-lining can reduce aerodynamic drag, the phenomenon still gets one's attention.

(Aircraft, too, have to live with the drag problem—it goes up for them roughly as the square of the speed as well. But they work much harder at streamlining and fly at high altitudes where the air is thinner, and the drag is thus significantly reduced. Getting thinner air is the motivation for the various schemes to evacuate the air out of tunnels for high-speed trains.)

The cost of energy is not the only issue here; there are other concerns about the availability of carbon-based fuels over the long term and the impact of their use on the atmosphere.

When all the potential sources around the globe are considered and the large reserves of natural gas and heavy tars and shales are taken into account, it's easy to say the world has plenty for years to come. The issue may not be so much that of running out of oil, the question is the price we might have to pay. The need for caution and prudence was discussed in Chapter 13 in the context of the Earth-warming issue.

On the plus side the "fixed" type guideway system will almost surely use electric power, just because this obviates the need to carry the extra weight of on-board fuel. This ability to run on electric power is a real advantage over highway-type systems that have much more difficulty in picking up the power from the roadway. While a lot of electricity is derived from fossil fuels today, electricity can also be generated by many other techniques. If the world has to shift to, say, more nuclear generation, the change is seamless to the system that already operates on electricity.

So the uncertainties in energy availability and the issue of global warming hang over our other decisions. As time passes, we may come to learn that all these concerns are overblown. We might find new sources for our energy or even decide that we really want more carbon dioxide to improve crop yields or to prevent global cooling. But we have to think about them.

References

[1] Federal Railroad Administration, *Assessment of the Potential for Magnetic Levitation in the United States*, Washington, DC: U.S. Department of Transportation, June 1990.

[2] Laithwaite, E. R. (ed.), *Transport Without Wheels*, Boulder, CO: Westview Press, 1977.

[3] Geerlings, H., "The Rise and Fall of New Technologies: Maglev as Technological Substitution?" *Transportation Planning and Technology*, Vol. 21, 1998, pp. 263–286.

[4] Money, L., "The Saga of Maglev," *Transportation Research*, Vol. 18A, No. 4, 1984, pp. 333–341.

20

A Superspeed Highway

The search is not for new ideas. It's for new ways to combine old ones.

We just explored the possibilities for a brand-new, high-speed ground system based broadly on the system format of the railroads. There may be another option: a brand-new, high-speed system based on the system format of the highway. In this chapter we lay out our thoughts on this second alternative, a new "superspeed" highway.

Recall that we started this whole discussion a few chapters back worrying about the safety problems caused by trucks and cars sharing the same highways. After looking at some possible ways to alleviate this problem, we were persuaded that where the situation is acute we will, in fact, build more highways and move in the direction of separating the two kinds of traffic.

And as a natural matter of course we will rework highways that are already in place, sometimes to expand them, sometimes just to cure the ravages of age.

The thought that enters our mind is why not see if there is some way we can leverage these investments by not just replacing or adding more of the same but by also upgrading their performance. And the two dimensions of improvement we hypothesize are a much higher permissible speed—perhaps a rough doubling to 140 to 180 mph—and, second, the ability to truly separate cars from trucks.

This is a tall order. We admit we have a hard time envisioning the geometric layout of a highway complex that can do, or at least evolve to doing, both of these things without taking up half the countryside. But we have a few years to work the problem, and we have little doubt that clever people will think of ways to do it more efficiently than we can think of now. It doesn't all have to happen at the same time; the cost might be spread over a longer time period by making provisions for, but not building, all parts of the new highway system from the very beginning.

Either way, it is likely that the complete new superspeed highway will cut a wider swath than today's highways. And the acreage we take up will be partially compensated by the reduced need for more lanes of conventional freeway or arterials in the vicinity of our new facility. The net need for space would, of course, vary from location to location and facility to facility.

We do not know how much faster we can go on our existing highways and still depend on human skill to keep it acceptably safe; our intuition says not over maybe 80 to 100 mph. But with automation to enhance safety (we wouldn't even mention superspeed if we did not see automation in our future) we could go much faster, perhaps to the 140 to 180 mph we mentioned earlier. And we would expect to fully apply all our tricks of automation to traffic flow and, therefore, have much greater capacity per lane than today.

There's no doubt that higher speeds will require more sophisticated road design. Vehicles will have to be able to "see" through bad weather and over the crests of hills and around curves, which will require communication between the fixed infrastructure and the vehicle. We will need longer acceleration and deceleration lanes at off and on ramps. There will be a host of problems that have to be solved.

In time we may be talking about completely driverless vehicles. But now we assume we will have human "drivers," operating through the ever-vigilant sensors and fast reflexes of an automated system. It's driver choice whether it's hands-on or hands-off cruising.

Maybe we should call them "vehicle managers" rather than "drivers." Given widespread automation of the driving function—maybe two or three decades from now—we see no reason that these automated vehicles, with their human managers, can't travel more safely at 180 mph than vehicles do today at 70 mph, particularly on a highway specifically designed for such operation.

Designing vehicles capable of such speeds seems perfectly feasible. As we said, we have plenty of production cars today that could easily cruise at 100 mph and faster, and over the years there will be more and more vehicles designed for even higher speeds. As sufficient automation is introduced to be confident that such speeds are safe, and as the fleet evolves toward more and more vehicles with these high-speed capabilities, they will find increasing miles of highway segments designed to accommodate them.

There is a transition that would have to be managed. In the beginning we would find ourselves designing new highway segments for superspeed vehicles with few superspeed vehicles available to use them. We see no option but to design the new highway facility, recognizing that it will start life carrying mostly these slow, old-fashioned cars like we drive today. This allows the new infrastructure to be useful from the very beginning. As the vehicle fleet evolves toward more and more vehicles with high-speed capabilities, more and more of the new highways would be dedicated toward accommodating them: highways will have to be designed to not only segregate truck and personal vehicles but to be adaptable to a changing mix of speed capabilities.

Step-by-step evolution is the operative thought. Even if we wanted to, the idea of just suddenly putting a whole new network in place over the whole country is just not realistic. One, it's too massive a project. Two, there is a serious chicken and egg problem: nobody will build a highway that is useless until a new class of vehicles is introduced and purchased in sufficient quantities to fill it. And no one will produce such vehicles until there is almost certainty that the new superspeed highway is there to accommodate and exploit them. And the highway cannot be just a few miles, but must be thousands of miles; otherwise the market will be too small to motivate the investment in vehicles.

And third, no amount of study will suffice to make sure we really have got it right, that we are investing in a winner. There are not only alternative technologies to be tested, there is the response of real-world markets to the new services to be experienced. It will take time, but preferred designs and approaches will start to emerge and a few primary standards agreed to. Then networks can begin to evolve around them. Only some operational experience can provide the insight and verification that larger, more extensive investments are warranted.

It is not inconceivable that we could evolve some of our existing highways into the superspeed system, but there are real difficulties. Today's modern design highways are typically designed for 70 mph. It is certainly possible that they could be progressively upgraded over the next few decades for higher speeds, letting them creep up from today's 60- to 70-mph speeds to 80-, 90-, and 100-mph speeds as vehicle automation makes such speeds safe.

The fact that the vehicles that operate on the new high-speed highway-guideway will also be able to operate on the city streets, making door-to-door trips possible without transfers, is a very important advantage over the fixed guideway systems that we discussed. Eliminating the time lost in intermodal transfer implies a higher door-to-door average speed than the necessarily intermodal movement on fixed guideway systems with the same top design speed.

A second major advantage that we car owners see is on-demand readiness. We don't worry about "frequency of service"; when we are ready to go, we go.

We think that the primary market niche for the superspeed highway are the trips generally below 500 or so miles; the air system will still remain king for longer trips. Does this "niche" matter? Is saving an hour on 150-mile trips or 2 hr on 400-mile trips important? One of the authors knows from experience that taking 2 hr off a 400-mile drive he makes quite often would make him ecstatic (especially if he could read a book on the way), but how many people make habitual 400-mile trips? How many would if we had automated, superspeed capability? Would our lives be better? Would our trucking system be more productive? Or would we just deplore the fact that broadening our travel horizons implies an increase in energy use?

We don't know. We do know that in the past increasing speed has changed behavioral habits and enriched the lives of individuals. But we don't know how much is enough. We suspect it's like money: all we need is just a little bit more.

The old familiar problem

All our past experience tells us that if we make travel easier and faster and not a lot more expensive, we will get more of it. We may improve our quality of life, but we will use a lot more energy in the process. We've already discussed the issues and concerns this raises.

We do think there is an advantage in systems that can be driven by electricity rather than by petroleum. Today these are largely equivalent, since we make so much of our electricity from petroleum. But it doesn't have to be so; there are other ways to generate electricity. As we noted, the "fixed" type guideway system is relatively easy to adapt to the use of electric power because continuous direct contact can be maintained between a power "rail" on the guideway and the power pickup on the vehicle. As far as we can see now, this is much more technically difficult for highway vehicles—it appears highly probable that they will continue to be directly dependent on petroleum for at least several more decades.

Today the dominant power source for highway vehicles is the internal combustion engine, a petroleum burner that becomes a source of both air pollutants and carbon dioxide. The former have been drastically reduced over the last decades, however the burning of any carbon-based fuel produces carbon dioxide, which is of concern because of the greenhouse effect that is arguably increasing global temperatures.

It is not out of the question that a replacement for the internal combustion engine will be found; in fact it seems likely. It is most relevant for the superspeed highway option, where our time horizon is two or three decades, since it will take that long for the automation necessary for higher speeds to obtain in a significant proportion of the vehicle fleet. It's not unlikely that in automation's latter stages of introduction it could also be accompanied by new power systems. There are already on-going developments in alternative power plants that may well fill the bill within that time frame, or substantially before.

The primary candidate now is the fuel cell. As we stated in Chapter 13, there is very serious work ongoing to develop hydrogen-air fuel cells to replace the internal combustion engine in the long-term future. It seems highly probable that for at least a few more decades this hydrogen will be produced by deriving it from some kind of petroleum. While the output of the fuel cell itself is just water, H_2O, the production of the hydrogen from the primary fuel will still produce some carbon dioxide.

But deriving the hydrogen from some kind of hydrocarbon is not inevitable: hydrogen can be produced using electricity for the disassociation of water. If we can generate that electricity from the sun, or using nuclear fusion or fission, then we can break the dependence of highway vehicles on petroleum. This would put an entirely new complexion on the whole issue.

We are pessimistic about the prospect of powering most vehicles with batteries. Batteries are very heavy for the energy they can carry because they have to carry it all—they can't get part of it out of the air as they go along as we do with our gasoline- and diesel-powered vehicles or as we will when we use hydrogen-air fuel cells.

Fixed guideway or high-speed highway?

There is a good chance that both have a place in our future; both certainly have a real advantage over either today's highway systems or our typical rail systems. And while the high-speed highway is the more flexible, the fixed guideway can support much higher speeds. There are probably some very large niches for both.

The fixed guideway approach—maglev is here the leading contender in our minds—fits best where we want to connect stations at which activities are highly concentrated and where the pick-up and delivery function is easy and fast. The example we noted is to allow several cities to better share one airport; an alternative is to integrate multiple airports into a large urban area so that they can essentially function as one.

The high-speed highway is probably the better choice for connecting larger urban areas where shipments or passengers might start anywhere in a many mile radius and go to similarly diffuse destinations. Here the ability to use city streets would be a very large advantage.

Are there other ways to go that we haven't considered here? Yes, there are, some we can identify now, and probably many that we haven't thought of at all.

What do we propose?

We propose serious consideration of all these possibilities. We encourage experimental and semiexperimental systems of all kinds.

We think the leverage of improved transportation performance on the world economy and the quality of our lives is much larger than just immediate market capture. It is worth investment of both brains and money.

The Venusian turned to the Martian and said, *"I like the way they are thinking now and am curious to see what they come up with. I was really worried that they would rush out and start building new highways without thinking through all their options. They are lucky to have had us around to straighten them out."*

21

Big and Slow, but Cheap!

We have focused on faster and better and neglected aspects of cheaper. So what happens when we focus on them?

Want a system with lower costs? We can think of at least four ways to start. The first route is to lower labor costs through automation. We are now and will continue to increase the use of automation in our existing systems over time anyway, so we don't have to build something entirely new to be taking advantage of this possibility. For example, advanced control systems are under development to increase the efficiency of truck routing and scheduling, and both truck and rail firms are adopting cargo and engine monitoring systems as well as location monitoring and advanced message communication.

The second path is to lower energy costs by going slower—the opposite tack from our conjectures so far. When we somewhat cavalierly decided to go faster, did that automatically imply higher costs? Not

241

necessarily; as we pointed out the impact of speed on costs really is a two-edged sword, largely the tradeoffs among vehicle productivity, vehicle costs, energy consumption, and other operating costs. If we design for slow, the vehicles are not expensive, and crew costs are reduced through automation, then the energy savings could be more important than the penalty on vehicle productivity.

The third possibility includes the exploitation of new, lighter, and stronger materials and the new approaches to design that they might make possible. Aluminum is already used in the cars of large, heavy coal trains hauled by advanced technology locomotives.

The fourth possibility is to ship in even larger quantities to get still better economies of scale, further extending the same trend that has led us to supertankers, 747s, and 100-car trains. As we already pointed out in the context of our small batch movement discussions, other things equal, bigger vehicles are cheaper per ton-mile because the payload weight becomes a larger and larger proportion of the total weight that has to be moved and because the costs of the power systems and the crew are spread over a larger base. A somewhat similar line of reasoning argues for larger docks, tunnels, and more lanes on freeways.

This last possibility has to be looked at carefully, however. The things we are transporting, like grain, don't all start at the same geographic point, ready to be loaded on some giant conveyance for transportation. They start as thin flows that are coalesced to produce the large streams. There may be more room for cost reduction in the collection and coalescing part of the process than in the "trunk line."

So we probably can build a new big, slow system that is cheaper than anything we've got today except barges on the Lower Mississippi River, where nature has supplied most of the rights-of-way and some of the motive power. (By a laughable coincidence, both the river and the barges are big and slow. Pipelines also have big and slow characteristics, but they are outside the scope of our discussion. They compete well with rail for the bulk movement of flowable commodities.)

The real issue is will anybody want to use our new system? In earlier chapters we talked a lot about the influence of size on service and noted the tradeoff between size of shipment and frequency of service. We can conclude that if we design for big and slow, service in terms of speed of delivery will be terrible.

As we see it now, the only potential market for a new big-and-slow system is for bulk commodities like iron ore, coal, grain, wood pulp, chemicals, or bauxite. These are important: the flows really are huge and, when such large movements are involved, even small unit-cost savings add to large numbers. "Big and slow" would not be competitive for higher value goods movements; these want faster service. And while high value, small batch is the fastest growing goods movement market, it is not the market that our new, cheap system will serve.

The dance goes on

"Seems to me you are being rather myopic," says the Martian, a claim he makes rather frequently. *"It may seem hundreds of years ago to you, but from my view, it was just yesterday when Portuguese, Spanish, Dutch, and then English merchant-men increased the sizes of their wooden ships and improved material handling at docks in order to lower and lower the costs of moving furs, products of the Orient, tobacco, cotton, sugar, and later wheat. It was also just yesterday when Brunel introduced big ships. Ships then got bigger and more specialized to markets.*

"Not only did this result in discovering that new things became eligible for this kind of transport, but that whole new opportunities for new routes and new uses opened up—the second step in this two-step dance you keep talking about.

"I know you are not into the ocean trades in this book, but drawing on lessons from ocean shipping as well as lessons on the land should add more depth to your thinking."

We are again challenged to be more imaginative. We start by recalling that cheaper has meant that market and raw material sourcing and product distribution areas can be increased. As a result, both production and consumption may involve a greater variety of products and services. Variety and specialization go hand in hand and may offer opportunities for transportation-based innovations. Cheaper can make us better off in ways that go beyond a penny saved here and a penny saved there.

Until now we have mused about a lower cost system, a service that is pretty much ubiquitous and that competes with and compliments the present rail and highway modes. But we hear from many people that we have already done all that can be done. People are always trying to squeeze out the last penny and can quite reasonably wonder if we haven't gone about as far as we can go (everything is up to date in Kansas City?). Most tradeoffs don't seem to favor lowering costs by going slower. Also, equipment-based scale economies come hard because of limits on vehicle sizes if equipment is to operate system-wide. It's easy to reach the not-optimistic conclusion that there is no more blood in this turnip.

The Martian just can't keep quiet. *"That is exactly what folk were saying about ocean transport toward the end of the wooden ship era and again in the late 1940s when just about every trick to lower costs had been tried."* And Martian is correct. Indeed, Texas economist Eric Zimmermann wrote a well-reasoned book in the 1920s explaining that because of competitive pressures, ocean shipping was optimized on every dimension—there wasn't much more that could be done to improve it [1].

Due to Zimmermann and the opinions of many others not withstanding, there have been enormous changes in ocean shipping since the 1940s [2]. Do these changes offer hints about market niches?

As we have seen over and over again, the words specialization and innovation apply, and specialization applies to routes, ships, and ports. That's what one sees in the coal, wheat, petroleum, container, and other trades. There is also getting the scale right and evolving appropriate port and ship technology. Appropriate scheduling, financing, and other control and institutional arrangements were also needed.

With that reminder in mind, what do we see when we look away from big existing systems to market niches? We see lots of possibilities; some examples will illustrate these.

More and more ships are going into service that are too large to traverse the Panama Canal—locks on the canal are too narrow and too shallow for the larger, more efficient ships appearing in many trades. There are lots of other too small, too slow, or too something "chokes." The St. Lawrence Seaway situation at larger geographic scale (connecting the Great Lakes to the Atlantic) has those characteristics.

And there are barriers on land, such as those formed by the Swiss Alps, the Appalachian Mountains, and by mountain ranges affecting routes in and out of California. The configuration of the Great Lakes blocks easy transport between the eastern and western provinces of Canada. Routes must go north around Lake Superior or south of the Lakes and through Chicago.

For a regional scale example, consider the haulage of grain from farms to country silos and then on to larger silos. This is currently accomplished using trucks or farm tractors hauling wagons. There are a bunch of problems, including the high cost of truck operations, road damage, and the cost of maintaining an extensive rural road system.

Looking at urban areas we see the movements of building materials, fuels, waste, and other things in situations where traffic conditions and roadway designs constrain the services that can be offered and thwart achieving lower costs.

These are examples of niches where doing old things in new ways might be in order. What about new things? Create a portable water supply by moving ice from the Antarctic; collect and move logs in order to save the lumber from trees cut in urban areas; dredge top soil collected behind dams and move it to renew soil on farms; and establish low-cost, temporary water and fertilizer systems to serve a year or so as semiarid to arid areas are reforested.

Some day the desalination of sea water will become an important source for dryer climates. Then transport from the seacoasts to the arid inlands might dwarf any bulk movement we've seen to date. While pipes and canals are likely to dominate this movement, we can imagine situations where overland transport could well become important. The list goes on.

Suggestions

We have glanced away from places where the railroads and other modes have honed and honed cheaper services to look for niches that open possibilities. We gave some examples, but far from exhausted illustrations of possibilities.

Turning to suggestions, let's look at the tried and true, the same sorts of things we looked at in the first paragraphs of this chapter. We take another look at them.

Automation? We have heard of ideas for automating the collection of garbage and other waste and bringing it to truck pick-up points. The old-fashioned garbage-trash chute in an apartment building perhaps qualifies for automated movement, and there have been trials of collecting from sets of buildings using pipes pumped to vacuums. Mail is distributed by automated carts in business offices. Make that slow system bigger to serve downtowns and include parcels of all types?

Scale opportunities? A study of the transport of grain to elevators suggests that large trucks with two or more times the hauling capacity of those used today would lower total costs [3]. (Technically, lower the joint costs of equipment, operations, and road maintaining. Cost would be minimized by large trucks operating at low speeds on roads of light construction.) Rather than trucks, use self-unloading trains to move sand, gravel, and other building materials. Improve heavy-haul technologies for the movement of preconstructed housing.

Economies of scale could also help overcome barriers. Returning to the Panama Canal bottleneck, reconstructing the existing railroad or building a new one might take advantage of scale economies. One might think of a railroad with a gauge of, say 18 ft, so that containers could be placed and stacked side by side. We imagine "land ships" connecting Pacific and Gulf ports.

While we are at it, let's think of combining automation, larger scale, and evolving sensor and information technologies with endless belts for ores and coal, and pneumatic systems for grain, for example. Looking beyond Panama, let's look at the Suez Canal and ways to access interior Africa and Asia, for example. Let's think about a variety of services in a variety of places. Variety and combining are the operative words.

Why not more blood from those turnips?

The Martian has a questioning look on his face. He says, *"I'm having a problem with your discussion. You've given us lots of possibilities for reducing costs. So why aren't these possibilities already being taken up or been mined out as you suggested before?"*

This is question that makes us think, and we think situation matters. For example, we said that barriers pose possibilities. Suppose a mountain range forms a barrier. That's no problem. If a railroad desires, it could take actions to reduce costs, say, ease grades and curves and build tunnels. An agency could build a freeway or toll road for large truck traffic. Nothing to be concerned about here. Improvements will occur as traffic warrants and as new developments, such as improved tunneling machines, permit.

The Panama Canal situation is different. There are several actors: the existing railroad, maritime liner operators, and the Canal that might lose traffic to jumbo rail services. Importantly, there are the actors that would be involved in port operations, financing, and obtaining government approvals, for example. In situations such as this, making markets for improvements may be costly and chancy. Perhaps progress will come as we innovate ways to involve actors with conflicting agendas and learn more about undertaking costly and chancy endeavors.

That sounds familiar. It is a comment that holds for many of today's proposed investments. Perhaps it is faulty and limiting to think of barriers as mountains and other physical things. The barrier is our inability to get players together and appropriately divide costs, returns, and risk. Imagination and energy are need as well as the discipline required to avoid unworthy politically motivated developments.

With a little help from our friends

"I've been quiet," says the Venusian, *"because Martian has been pushing you pretty hard. But I think you need more pushing.*

"Several chapters ago, you gave us that nice history lesson involving Pease and Stephenson, and two parts of that lesson spill over into what you are saying. First, you seem to say that the costs on existing systems have been ground down and down and things are pretty wonderful. That is just what Pease and Stephenson were told: the roads, tramways, and canals of the early 1800s were just fine. Second, while not the same, your situations are cousins of the Stockton and Darlington situation. They are tough places to serve.

"You have concentrated on Stockton and Darlington-like market niches—to repeat, tough places to serve. But there are other kinds of niches. There is more to think about."

"What you have said is too vague and general. Keep trying and you may get it," says the Martian. With his maritime hat still on, he says, *"I'll give you another hint. Recall what I said about the wooden ships—how they were increased in size to lower the costs of moving tobacco, sugar, and such. For instance, the big ships in the India trades (Indiamen) could haul as much as 500 tons. The*

measurement protocols differ, but that is roughly about what can be moved in four of today's gondola rail cars.

"And think about what happened then and has happened recently. Many products that used to be shipped in small batches at high cost now move in bulk (neobulk) in specialized ships: ordinary wine and liquefied natural gas in tankships and automobiles and steel in specialized ships are examples."

Full circle

In the previous chapter, we concentrated on products for which "time is (lots of) money" and asked how high-quality small batch services could be improved. In this chapter we began by concentrating on cheaper, and our thinking led us to large scale bulk movements such as those represented by coal and grains. We began to think slower and cheaper. (No analogy intended.)

In-between commodities fell through the cracks. Pushed by our friends, we now remember that cheaper is relative. With the growth of population and trade there is more freight being moved. There are opportunities to be grasped as medium-sized batches take on bulk-like movement characteristics and as individual shipments act more like batch shipments.

Cheaper is one route to variety and the new production and consumption options that transportation improvements may offer. With that motive for lower cost in mind, the future of lower cost services should be very good.

References

[1] Zimmermann, E. W., *Zimmermann on Ocean Shipping*, New York: Prentice-Hall, Inc., 1921.

[2] Gardiner, R., (ed.) *The Shipping Revolution: The Modern Merchant Ship*, London: Conway Maritime Press, 1992.

[3] Fawaz Y. M., and W. L. Garrison, "Truck and Highway Combinations for Increasing Trucking Productivity in Market Niches," *Transportation Research Record*, No. 1430, 1994, pp. 10–18.

Part V

The Air Transportation System

22

To Grandmother's House We Fly

"Heavier-than-air flying machines are impossible."

Lord Kelvin, *President, Royal Society, 1895*

If I had my druthers someone would pick me and my baggage up at my house, preferably in a limousine equipped with libations and maybe TV, and deliver me directly to the door of the plane five minutes before takeoff. Naturally I would expect comparable service to get me from the plane to Grandmother's at the other end of the trip.

I would, of course, fly first class (at least).

What a lovely thought for the future. The only improvement we can think of is to speed up the trip to the airport. Invoking Scotty for direct beaming makes us a bit nervous, so maybe speeding up the limousine to, say, 180 mph would be good enough for now. This would cut a 30- or so mile trip to the airport to 10 min.

How closely is it feasible to approach that super service we just described? (We assume none of us is President or own our own jets.)

The short answer is that most of this is a pretty tall order. Perhaps we are suffering from a failure of imagination, but there are a lot of practical—and economic—difficulties in the road to that kind of service. But we have been told that a man's (person's) reach should exceed his (or her) grasp, or what's a heaven for.

First, it seems totally impractical today or in the foreseeable future to have cars all over the airport delivering and picking up passengers in ones and twos directly and from the various planes. The inordinate time required for the loading and unloading operation, the baggage-sorting problem from arriving planes, and just the sheer volume of vehicular traffic seem too daunting. Just handling the on-the-ground traffic of the aircraft themselves is no small job at a large airport.

So it's hard to imagine how we can avoid having some facility that is physically separate from the aircraft themselves where passengers and baggage are consolidated into bigger batches for relatively prompt enplaning and deplaning.

I think we've just invented an airport terminal.

In years past at the Dulles Airport in Washington, DC the passengers were delivered from the terminal to the planes in large, specially designed "buses"—they call them mobile lounges—that fit directly onto the planes' doors; these are shown in Figure 22.1. This technique permitted a smaller terminal to still service a large number of aircraft because it was not necessary to provide a lot of space for an aircraft next to the terminal in order to load and unload, just enough space for a bus bay. The Dulles terminal is long and narrow, so the trip from the street where one might be dropped off to the airport side where the special buses were was a relatively short walk.

Passengers had mixed reactions to this arrangement. Many appreciated the less-sprawling terminal and the shorter walking distances, but others would have preferred walking directly to an aircraft gate even if it meant walking farther rather than spending the extra time and modest hassle of an intermediate "bus" trip between terminal and aircraft.

The Dulles terminal has now been overtaken—and overwhelmed— by the growth in air traffic. Today a good part of the original building is taken up with the usual airport and airline services and the security check

Figure 22.1   The Dulles International terminal of Washington DC, showing mobile lounges. (*Source:* Metropolitan Washington Airport Authority.)

area. The mobile lounges now go directly to only a few of the aircraft; most now take all the passengers to a second, more conventional terminal building where all the aircraft gates are located. Today the vehicles are largely nothing but substitutes for a long tunnel to access this newer, loading-and-unloading terminal. Now most passengers get double the pleasure: both the trip on the mobile lounge and a walk to their aircraft gate.

Nearly all airports are now designed for the aircraft to park immediately next to the terminal, so the walk to the aircraft is usually a short trip through a moveable tunnel running from the door in the terminal—the "gate"—to the door in the aircraft.

Requiring that the aircraft be directly accessed from the terminal gate implies that the span of terminal required for each gate is dictated by the wingspan of the largest aircraft that will be parked outside—plus a little maneuvering room. Making the terminal curved helps a little, because the aircraft are farther out on the radius than the terminal itself.

Terminals have thus become long rows of gates laid out in various geometric arrangements. Thus the distance between Gate 14 and Gate 24 has been dictated by the need to fit space for nine aircraft between the two, not by what an air traveler considers a reasonable walking distance. All our larger terminals, such as the Dallas–Ft. Worth International Airport shown in Figure 22.2, have internal transportation systems—like moving walkways, carts, or little trains—just to help the walking-averse passenger get between gates, between subterminals or "concourses," to baggage areas, to street-side transportation, to restaurants and stores, and whatever.

Thus, the decision to build significantly larger aircraft than those that the taxiways and terminal geometry is now designed to handle will rattle through the whole system, forcing modification or replacement of existing terminal buildings and perhaps the taxiway layouts. One way around the problem is to require that larger aircraft have wings that can fold up, but this, too, carries a cost. Another alternative is to emulate the mobile lounge approach used at Dulles. (A glance back at Figure 22.1 lets one visualize the difference in the demands placed on terminal geometry.)

Figure 22.2 An aerial view of some of the terminals at Dallas–Ft. Worth Airport showing aircraft parked next to the gates. (*Source:* DFW International Airport.)

Urban terminals

A person unencumbered by too much knowledge about airports might make the following suggestion. *"If we are going to use bus-like 'mobile lounges' to go to the airplanes, why can't we use redesigned versions of those same vehicles to help passengers get to the airport? Instead of just running back and forth between the terminal and the airplanes, they could start from some convenient location nearer town and then go straight to the airplane. Why not have little check-in 'terminals' all around the urban area, where one can board a bus that then goes straight to the airplane?*

Rather than have one big terminal at the airport as we do today, break it up into a few smaller ones scattered around the urban area. Then an air passenger need not get all the way to the airport, but only to the urban terminal nearest them, which would be a much shorter trip. Passengers using the urban mini-terminal would presumably find more convenient parking and at-the-terminal rental cars. They would check their baggage there, go through the necessary security screens, and then be off directly to their airplane on their bus."

This may be overkill of a basically interesting approach. First, there is the problem of building a bus that can operate on ordinary streets and highways and still serve the role of a mobile lounge, raising itself in the air enough to fit against the door of modern aircraft. Second, there is the problem of direct delivery we mentioned earlier: with several urban terminals serving multiple flights, there would be far too many vehicles running around the aircraft.

And even with urban mini-terminals, we will still need a terminal at the airport, if for no other reason than to serve the passengers who are just transferring to other flights—they don't want to have to go all the way into town to get from Flight 423 to Flight 27.

The whole idea looks much better if we just skip the part about going directly to the airplanes. Now we can use ordinary buses to go to the airport terminal, not the aircraft themselves. We can run the passengers through their security check at the urban terminal and then unload them and their already checked baggage into the secure side of this airport terminal. Then these already-checked-in and screened passengers only have to walk to their aircraft's gate. They have still avoided a lot of the hassle of a big airport and having to cope with the long drive out to it.

There may be merit in this scheme at our larger airports. It effectively puts the check-in, people/baggage-processing "terminal" much nearer to where folks' trips actually start and stop. It might make parking a little easier and maybe eliminate the extra bus trip usually needed to get back and forth to a car rental lot.

Like most ideas, this one is not new. Many years back, a similar scheme operated from downtown Washington, serving both Dulles and the Baltimore-Washington Friendship airport. The Eastside Airline Terminal in Manhattan operated from 1953 to 1973 [1]. At the Eastside Terminal there was passenger check-in and buses were scheduled for specific flights. But buses began to operate on schedules unrelated to flights and the facility closed soon after airlines began to eliminate check-in services.

Overall, on-call van services and other arrangements have provided strong competition to bus services, and for that and other reasons off-airport terminals have had limited roles. Limousine pick-up is sometimes offered as a service to first-class passengers. Tour operators bring buses to customers at hotels. There are lots of ideas, and varieties of services may bloom as airlines and travel brokers strive to specialize services.

But separating the people-processing facilities from the on-airport terminal would cut down on the services needed at the on-airport terminal. Because the on-airport terminal could more easily be divided into multiple, physically separated structures, perhaps a bit more efficient layout of the runways and taxiways would be possible, improving operations from the aircraft point of view. Some transport between structures would be needed to serve the few hubbing passengers not leaving from their arrival structure.

But one thing urban terminals probably wouldn't do is save time on the trip to the airport—they would use essentially the same roads the passenger would have otherwise used. Saving time is no trivial matter: most of our larger airports are already well out on the fringes of metropolitan areas, and any new ones are likely to move even farther out.

Is getting to the airport a sensible application for one of the new high-speed grounds systems, maybe a magnetically levitated system—maglev—as we discussed in Chapter 19? (Our superspeed highway is probably a bit farther in the future but is still a candidate.) Maybe maglev does make sense. Maybe it will be the keystone that permits a major rethinking about how airports are organized. We will have much to say about this later.

Terminal interiors

Although we haven't discussed it, we all know of increased choices as airlines offer different classes of services at different prices and with different restrictions. Some airlines focus on low costs with skimpier services. Local service, regional, or feeder airlines have their specialized market niches. Service differences among airports are found in cities where there are multiple airports. Finally, there are a variety of ways to get to and from airports—use and park a private car; get dropped off by friends or family members; take a limo, van, bus, or occasionally train.

Inside airports we find all kinds of shops, bookstores, and other varied sales and services. Yet there is a certain lack of variation. The gate used for tourists to Florida looks just like the gate serving mainly business passengers. Grandmother arriving on her first flight exits the plane into a stark, unfriendly corridor along with seasoned passengers who know where they are going. Grandmother's needs are different.

The family sending Charlene east to college sits in a row of chairs across from folk on a tour. What a horrible seating arrangement. It forces telling the baby good-bye while staring at strangers in the row across. We are saying that one-design-fits-all airports may not serve very well when passengers have wildly different information needs, skills, experiences, emotional situations, and other characteristics.

Think about it. Airport designs do not serve many social situations such as families greeting loved ones or saying good-bye very well. Information is often confusing. Which way does one go to get a rental car?

Ours are not new observations. Written some years ago, a book on the social history of air travel introduced the era of widely available services circa 1955 using the phrase "passengers as walking freight" [2]. Have you ever been in a truck freight terminal? Shipments are jammed close to gates waiting to be loaded. There is noise, uncertainty about when the truck will arrive, conflicts between arriving and departing shipments, and more.

The airport designer says that efficiency drives the designs. With airline and airport managers counting pennies, the least expensive, standard design is expected. It's not surprising if we consider the carrots and sticks that drive behavior: airport managers almost have to be more concerned with shops where there are rents to be collected than with profitless waiting areas. Even so, it should be possible to do better.

Modern developments in the control of lighting, moveable furniture, providing information, and lots of other things could well differentiate services and environments for service users in tomorrow's terminals. The lounge just occupied by button-down business types might be relighted and otherwise reconfigured to send the team off for world soccer matches.

We can help Grandmother—and lots of other people, too—in other ways. Many of us have used receivers and earphones in museums. As we approach a painting, we hear about the artist, style, and more. Grandmother arriving in her first flight could be given something like that—reassuring, specialized guidance in the language of choice. "Turn left to go to baggage claim," "entrance to customs is straight ahead," "baggage from your flight will be along in five minutes." We used to talk about the "talking ticket" but hardly know what to say with tickets rapidly disappearing.

Depending on competition, the motivation to spur the more friendly terminal may be misplaced because the extra investment doesn't provide an advantage for any one particular airline. And it's probably hard to prove that the better terminal increases net-demand enough to spur the airlines to collectively undertake such improvements. It appears to us that if it is to happen, it will probably require some other impetus—like the local airport authority—to spearhead the move.

Around the airport

Airports grew on green fields. Inexpensive land was needed, so rural land at the edge of the city was acquired and developed. As cities have grown many airports have been converted to urban uses, and other airports have had their growth constrained. They have been overwhelmed, swallowed up, or contained by urban growth.

Just the reverse is the case for the dozen or so really big air traffic hubs. Those airports have strongly impacted and sometimes overwhelmed their environments. There is a lot of airport-related traffic, and there are off-airport parking facilities. There are also hotel and meeting complexes, office buildings, and on- and off-airport freight-handling facilities. Often there are apartments catering to airport and airline employees.

But in both cases, the external growth inhibits internal expansion. The noise and emissions and the congestion that the airports generate are not happily compatible with any kind of activity outside its boundaries and naturally create a resistance to allowing any of these from getting worse.

The airside

Our discussion so far has focused on the so-called groundside of airport operations—access to airports and within-terminal activities. Airports also are concerned with the number of airside operations—landings and takeoffs—they can support. Safety requires that aircraft be kept to reasonably large separations in the air and that the runways are clear of the prior operation before a new landing or takeoff. This is a point that is perfectly obvious to all of us who have waited in a long line of aircraft waiting our turn to takeoff. These constraints obviously limit the numbers of landings and takeoffs that can be supported daily. Uncooperative weather doesn't help the problem.

There are all kinds of air traffic control tricks being developed to try to increase the number of aircraft per hour that can be accommodated on one runway. And there are some gains still to be made along these lines.

But the fundamental way to increase an airport's airside capacity is to add more runways. This is not easy. It takes more real estate that many older airports, already surrounded by development, can't get. As we noted, airport neighbors are concerned about increased air traffic and noise and congestion. Increasing airside capacity with lots of runways thus becomes the prime motivation for new airports way out of town where lots of land is available.

And this is one reason we talk about airports 50 and 100 miles out of town in the future. The other is, of course, that airports don't make particularly good neighbors, and no amount of good fences can really fix that problem.

Do we need to have all our new runways at the same airport? Instead of four runways on one airport, why not two each on two airports, built closer into town?

The problem is at the larger hubs where one half or more of the passengers are there just to transfer to another plane [3]. If a plane is at the other airport, the transfer passenger views it as more than a slight hitch in the trip and vows that next time Amalgamated Airlines will be chosen so that the connecting flight is at the same airport.

We should mention that in many ways air freight is analogous to people. Much of it travels in the bellies of passenger planes and so is inseparable from passengers, and much of it is also slated for transfer to other planes to reach its final destination.

Why not more direct flights so that it doesn't matter which airport? We're back to the small batch problem. Many cities don't generate enough traffic to support lots of direct flights to and from lots of destinations, so passengers (and freight) are flown to a hub and redistributed to their final destination.

It's true that as overall traffic grows there will be room for more point-to-point flights. But growth also asks for more hub services. And, note, if we did have lots of smaller aircraft economical enough to compete with point-to-point flights, we may have increased the airside operations and thus the strain at some airports to accommodate them all.

It is a complex system, and there is just one tradeoff after another. Strategic thinking is needed [4].

The airport system of the future?

Our friendly critic who doesn't yet know that things can't always be done asks, *"Why can't we put a very high speed link between the two airports so that the transferring passenger can get from one to another in 5 or 10 min; they shouldn't mind that too much. That's no worse than getting from one gate to another in Chicago or Dallas–Fort Worth or Atlanta or Houston."*

The fact is that this appears to us to be a very sensible application for a high-speed link between airports: it enables a different way to think about airport design—what we would call a new design pathway. Maybe we have just invented the integrated airport with spatially distributed runways.

We've already talked about the spatially distributed groundside, with multiple mini-terminals spread around the urban area, connected by high-speed transportation systems to perhaps spatially distributed-on-the-airport gate facilities. Now we've added the option of distributing the runways themselves to different locations and integrating them from the perspective of the passenger with high-speed ground links. These may be maglev, or some form of a minisuperspeed highway with its automated vehicles, or something else we haven't thought of.

A new high-speed ground system would give us a lot to think about for the design of new airports and increasing the usefulness and capacity of the existing ones. Indeed, a proposal for an O'Hare–Milwaukee Airport Maglev connection has been floated off and on. Why not extend it to Madison. An O'Hare–Milwaukee-Madison Airport is something to think about—but for tomorrow, not today.

We may never fully hit our ideal air trip, but we may be able to better cope with future growth and still get to Grandmother's house on time.

Faster airplanes?

There's nothing like a trip to the Far East to make one think about supersonic transports—like the Concorde, only bigger and faster. The possibility of building such an aircraft was very extensively studied in the 1960s and ultimately rejected, primarily on environmental and energy-consumption grounds—fast aircraft require more fuel per mile than slower aircraft. This supersonic transport, the SST, was primarily aimed at service over the Atlantic, replacing the 6- to 16-hr trip times of subsonic aircraft with something like 2- to 7-hr transit times. The Organization of Petroleum Exporting Countries (OPEC) was the stake through its heart.

The specter of the sonic boom didn't help. For those who have not been exposed to such things, an aircraft flying above the speed of sound generates a so-called shock wave in front of it. These shock waves are like a thin, almost semisolid wall of compressed air that, pushed by the aircraft, is also traveling faster than the speed of sound. This shock wave forms a big, irregular cone off the aircraft's nose, and the lower part of the cone sweeps along the ground. When this wall of compressed air sweeps past, people and animals hear it as a loud "boom," and the wave of air exerts appreciable pressure on the objects that it hits. While the possibility of actual damages to some structures is possible, shock waves are generally not dangerous but are loud enough to startle, even frighten. The Air Force has tested supersonic aircraft on their test ranges over the western United States and has bought many chickens that allegedly stopped laying eggs because of the sonic booms.

Because of the sonic boom there is no serious thought of operating any commercial aircraft at supersonic speeds over land.

There was a brief revival of interest in the early 1980s for a supersonic aircraft that could serve the much longer trips over the Pacific. There were many studies of various aspects of the possibility carried out at the time; the principal conclusion from most of them was that improving technology had made such an aircraft feasible and lowered seat-mile operating costs to within competitive range of the comparable costs for subsonic jets [5]. But the environmental and energy consumption issues had not disappeared, and perhaps there was even more concern about the acceptability of sonic booms, even over water. Another conclusion was that such a project would not only be risky but very, very expensive. Not surprisingly, nothing happened.

In addition to the shock wave off the nose, there is a counteracting shock wave also formed off the tail; there has been some thought that the two might be made to intersect and cancel each other below the aircraft so that nothing reaches the ground. There is also thought of reshaping the fuselage to break up a big sonic boom into many smaller ones [5]. The authors are not up-to-date on current thinking about the feasibility of such schemes or any other ways by which these shock waves might be made more neighbor-friendly. But unless something can be done about shock waves, we suspect that complaints from ships and islands would put more and more restrictions on the routes that could be flown supersonically even over the Pacific.

And the issue of greater fuel consumption won't go away, nor will the possibility of unacceptable pollution on the upper atmosphere; these aircraft will probably operate at 60,000 feet or so. Back in the 1960s and early 1970s considerable attention was given to this concern—and that concern, valid or otherwise, would surely loom again.

All these things add up to considerable headwinds for a new supersonic transport, at least for a while. But it surely would be nice to cut the 16-hr trip from San Francisco to Sidney to 7 hr.

Technology, congestion

So far, we have taken continued technological advances in air transportation as a matter of course and continued advances are expected. As a look at an industry magazine such as *Aviation Week & Space Technology* will reveal, there is excitement about new materials, gains in safety and energy efficiency, the tailoring of designs and services to market niches, and improved aircraft control. Advanced information systems are found throughout the industry. Automation and other uses of electronic and control technologies are expanding promising increased safety and efficiency [6]. There is also a lot of healthy debate about industry and government activities and the future of technology.

Excitement heightens at about the time of the summer Air Shows, a good time to watch the news for more.

Taking technology improvements as given, much of what we said previously in this chapter built on what everyone who has traveled very much knows: there is often congestion on runways, in the air, and within airports. Like congestion on the roads, congestion in the air system is caused by more aircraft and people wanting to use the system than there is easy capacity to handle. The problem is worse at some times of the day or when the weather is bad. And weather problems at some airports may affect flights and operations at distant airports.

We asked how to improve services while expanding capacity.

Early in this book, we talked about smart cars. Airplanes are certainly smart, and their airways and traffic control systems have smart features. They are getting smarter. We said that the smart car will help us increase the effective capacity of the road system. But not quite yet. What's the air transportation situation?

On the airside, gridlock (airlock?) has been avoided by using larger aircraft and increasing the number of runways. There is metering (flow control), especially on bad weather days when airport capacity is reduced.

There is a good bit of discussion of free flight—aircraft selecting the most advantageous from-here-to-there-route and avoiding lining up on often less desirable electronic airways. Enabled by smarter aircraft and traffic sensing and control systems, faster airport to airport, safer, and cheaper are potential advantages. Investments in capabilities based on communications, sensors, and location calculators will be required. But the payoff is thought to be great, so it will come. Already, approximations are found in cross-ocean flying. But the biggest problem is not in the air, it's at the airports.

Congestion pricing is significantly discussed—charging higher landing and takeoff fees when approach and departure capacities are congested. This idea is beginning to take hold, and we know that what happens at one airport may affect others. Mix these elements with developments in communications and airline flight economics, and the stage is set for rather sophisticated approaches to airport-access pricing [7].

For example, suppose that although we are recovering from thunderstorm-induced delays there will still be congestion delays for 4 hr at Our Airport—not enough landing slots for expected traffic. What is bid for the available slots? Ticket pricing and aircraft scheduling systems permit the airlines to know just what that slot is worth to them. An airplane where the passengers didn't pay for quality of service (higher prices) might sit on the ground at another airport while the plane with more expensive ticket holders gets better service.

While nobody really likes the idea, such pricing seems a natural progression for the kinds of ticket pricing now common (higher prices for higher quality services) to aircraft seat pricing.

Progress is also being made in making aircraft more energy efficient and better neighbors. Noise suppression is already widely adopted. Free flight should help energy efficiency, and fitting aircraft to passenger loads does too. A geared turbine is under development, and there is another 15% fuel economy possible there. But the bottom line for air transportation looks similar to that we have seen before: there are lots of things that help, but the growth of service presses hard on efficiency gains.

We have to run faster just to keep up.

Service improvements? Air taxi anyone? Commuter aircraft, corporate aircraft, small private plane? The statistics for commercial aviation tell us a lot, but there is this catch-all category of general aviation about which we know little. Not achieving the economy of scale of commercial aircraft in service between large airports tends to inflate their costs.

But service quality is also high. There is a cost-lowering innovation as aircraft are increasingly owned and operated on a time-share basis; the variety of business aircraft on the market may stimulate more and more specialized services. Innovation often takes place in the high end of the market and then spreads. But we can't afford to forget that more use of smaller aircraft means more landings and takeoffs to be accommodated.

It is wonderful to have the air system widening the variety of services it offers. Recreation, business, and other activities will surely continue to demand more services and we have far from exhausted the things we might do in response. But better service and cost choices are only the first step; remember our two-step dance. Perhaps even more important are the new kinds of activities and advantages made possible, activities that can enlarge our lives in ways that are now unimagined.

Flying on a reasonably large scale by the ordinary person is a post–World War II phenomenon, and many would argue that before 1980 it was largely limited to the businessman on an expense account rather than the ordinary person. The global air system, open to all and used by many, is now revolutionizing our view of the world, with long-run ramifications that defy our foresight. And the air system, in our judgment, is still only adolescent, with lots of learning and growing still in its—and our—future.

Selected bibliography

Horonjeff, R., and F. X. McKelvey, *Planning and Design of Airports*, 4th ed., New York: McGraw-Hill, 1994.

de Neufville, R., *Airport Systems Planning,* Cambridge, MA: MIT Press, 1976.

References

[1]   Gosling, G. D., "Airport Ground Access and Intermodal Interface," *Transportation Research Record*, No. 1600, 1997, pp. 10–17.

[2]   Hudson, K., *Air Travel: A Social History*, Bath, Somerset: Adams and Dart, 1972.

[3]   Morrison, S A., and C. Winston, *The Evolution of the Airline Industry*, Washington DC: The Brookings Institution Press, 1995, p. 44.

[4]   Caves, R. E., and G. D. Gosling, *Strategic Airport Planning*, New York: Pergamon, 1999.

[5]   Rosen, R., and L. J. Williams, "The Rebirth of the Supersonic Transport," *Technology Review*, Feb./March 1993, pp. 22–29.

[6]   Billings, C. E., *Aviation Automation: The Search for a Human-Centered Approach*, Mahwah, NJ: Lawrence Erlbaum Associates, 1997.

[7]   Oum, T. H., and Y. Zhang, "Airport Pricing: Congestion Tolls, Lumpy Investment, and Cost Recovery," *J. Public Investment*, Vol. 43, 1995, pp. 353–374.

23

The Los Angeles Airport System in 2020, 2040, 2060—a Parable

"There is nothing that clarifies issues more wonderfully than hindsight."

Chairman, Airport Planning Commission of 2020 A.D.

To date, we have largely treated transportation generically, without reference to specific locales. Here we depart from our usual practice and present a kind of case study of the Los Angeles Airport System.

We are doing this for two reasons. One, it lets us better illustrate the notions about future airports we presented in the last chapter and puts those ideas into a rough time scale. Second, Los Angeles is, at the time of this writing, going through a very serious examination and controversy about how to handle its need for airport capacity in the future, and it is our belief that some of the concepts presented here have relevance to its thinking and analysis.

Los Angeles is not alone with the problem of how to accommodate air traffic in the future. Throughout the nation there are other large city metroplexes and spatially proximate groupings of cities that will, over time, face similar problems imposed by a growing air system, an increasing dependence on air commerce, and a spreading urban population.

The problem is not trivial. Airports seem to carry the seeds of their own destruction, or, at least, curtailment. They are first built where there are few people and relatively cheap land is available. Then the jobs and commerce they create become magnets for more folks to move near them, and as time passes, these new neighbors cramp further growth by complaining of the noise and the traffic the airport generates. Airports are recognized necessities, but they are not viewed as good neighbors.

The situation in Los Angeles today is just one point in an evolution in demography and technology that has been in progress for a very long time and, as far as we can see now, will extend a very long time in the future. We plan to make several stops along this continuum as we present our thinking.

We begin our story roughly 70 years ago. Think first of yourself as a member of the Airport Planning Commission of 1928, which we only now invented in honor of this occasion.

The tyranny of the seminal decision

If the Airport Commission in 1928 had known what we know now, would the Los Angeles International Airport be where it is? Possibly not, but it seemed like a good idea at the time.

The land was flat and unobstructed, with good drainage and no neighbors (remember, it's 1928). It was a bit over two miles from the ocean, so there was only a little fog. (Well, maybe a little more than "little" sometimes; we still remember from the 1950s the super-smudge pots being burned along side the runway in an attempt to dispel it. We also remember the hundreds of rabbits that watched every plane's movements with apparently avid interest.)

Except for the fog, the only other objection to its location was that it was a bit farther than desired from the only steady customer of the infant airlines: the Los Angeles Post Office, some 55 minutes away by truck. That objection faded when Charles Lindbergh, still fresh from Paris, flew in there and commented that he thought it was a nice place for an airport [1].

One thing the decision makers of 1928 did not foresee was that in a few years a man named Whittle would invent the jet engine. This was a huge step along the way toward the Boeing 707, which entered operational use in 1957. They didn't know that this innovation was going to require a complete remodeling of their airport and generally revolutionize the whole air system. Ah, the frailty of foresight!

After the 707 nothing was ever the same again. The whole air system, then designed around the DC-3 and its bigger propeller-driven offspring, had to metamorphose to fit an entirely new type of aircraft. Airports needed much longer runways, they needed to be able to dispense a new kind of fuel, and they very quickly needed larger terminals and bigger passenger gates. The Air Traffic Control System had to change to handle aircraft that flew almost twice as high and considerably faster, aircraft that had different takeoff, climb-out, cruise, approach, and landing characteristics than propeller driven craft. And the airlines themselves had a lot of learning to do to adapt their practices to this new technological marvel, the jet aircraft.

For the Los Angeles Airport—LAX—this meant that the runways had to be extended, and the direction of least resistance was to the west. This brought the new airport boundary right up to the sand dunes that edged Santa Monica Bay (think Pacific Ocean). The prevailing winds were from the ocean, so the aircraft took off over these dunes.

The almost unbearable noise under the aircraft forced the eradication of the very attractive Playa del Sol beachfront residential development already in place along these dunes. All those houses have long since been moved away, but one can still see the old network of streets, now empty, with weeds growing in the cracks, the whole sealed off with a chain link fence. Sepulveda Boulevard, the original western boundary of the airport, was relegated to a tunnel under the new runways.

The jet meant more that just longer runways, it and the growth that it engendered meant virtually a new airport. A major remodeling during the 1960s featured a spectacular centerpiece theme restaurant that was surrounded by parking lots and nearly hidden by new buildings almost before it was completed. It still looks very impressive from the air.

Since then, with the growth in population and commerce and incomes in the region, with the new capability of jet aircraft, and finally with deregulation and generally lower ticket prices, air traffic has exploded. By 1998 LAX was handling over 60 million passengers and 2 million tons of cargo annually, the result of some 2000 landings and takeoffs per day (in the jargon—2000 air operations per day) [2].

The series of small cities stretching out from LA to the east—most of which were more like villages in 1928—had by the end of the century grown together to form almost one continuous city from the Pacific Ocean to San Bernadino, Riverside and Redlands—the region currently referred to as the Inland Empire. Aircraft generally approached from the east, so now the approach pattern of landing aircraft is over roughly 70 miles of almost continuous urban development.

There were other airports in the region, but at the end of the century LAX was still by far the dominant airport, handling roughly 75% of the total passenger and 78% of the freight activity for the Metroplex [3]. Growth was a continuing pressure, and most users of the airport through the 1970s, 1980s, and 1990s can't remember a time when construction and remodeling wasn't going on somewhere: a new international terminal, continuous expansions of existing terminals, parking structures, more internal circulation roads, new runways, and—expanding into the areas adjacent to the airport proper—more parking, rental car lots, and other support activities.

By the 1990s this pressure had led to a plan for a multibillion dollar revamping to hopefully smooth out rough spots that past growth had induced and to provide new capacity to cope with continuing growth of both passengers and freight into the future. An almost doubling of airport capacity from 1996 levels was envisioned [4].

Clearly, it would have been virtually impossible for the founders of the LAX to foresee the future that lay ahead for their little airport. We are reminded of the Committee established in the 1840s by the British Parliament to decide which gauge was best for the budding railroad industry. As we recounted in Chapter 14, they said that the 7-ft gauge used by Brunel was best since so much had been built at 4 ft 8½ in that they would stick with that. We can just imagine a committee reviewing whether LAX was in the right place before committing to the major remodeling brought on by the jet. The verdict would probably have been that maybe some place else might be better but that too many businesses and too many peoples' lives were geared to the current location and they weren't sure where they might move it anyway.

But even if the planners and policy makers at the end of the century had not thought the location of LAX was totally optimum, that's where it was, and barring major catastrophe, that's where it was going to stay for the indefinite future. A major portion of LA's industry and infrastructure had adapted to its location and its operation, and any precipitate shifts in its current role could very well cause significant dislocations in the Los Angeles economy.

Anyone trying to change things, to adapt to new circumstance, almost always has to deal with the legacy of history. The past all too often locks us into development pathways established long ago, the inescapable consequence of some often forgotten seminal decision. It is fashionable to extol the virtues of long-range planning, ostensibly to ensure we always select the optimum pathways for future developments. The most dangerous thing we can do is to kid ourselves into thinking that by such planning we can eliminate the surprises the future always has waiting for us. But while serious uncertainties were—and are—an unavoidable part of the planning game, we have to try.

The view from 2000

Here, in the blink of an eye, we will transfer our membership from the Airport Planning Commission of 1928 to our own honorary Airport Planning Commission of 2000. We will summarize as best we can the situation facing the policy and decision makers from that vantage point in time as they attempt to prepare for their future.

As we noted, in the late-1990s the LAX handled almost three-fourths of all the passengers flying into or out of the Greater Los Angeles Metroplex and an even greater proportion of air freight. The lion's share of the other passengers into the region was split between the already saturated John Wayne Airport some 36 miles by bird southeast of LAX in

Costa Mesa (10% of the total); the Ontario International Airport nearly 50 miles east and slightly north of LAX (nearly 9%); the Burbank-Glendale-Pasadena Airport about 18 miles almost due north of LAX (7%); and small dribs and drabs to a few others like Long Beach [5]. The location of these airports—and a few others—are shown in Figure 23.1, along with the major freeways and highways that feed them.

LAX appears to be bursting at the seams, and the central issue of the day is its proposed expansion, a major remodeling that is forecast to cost some $8 to $12 billion [6]. If this proposed remodeling of the Los Angeles

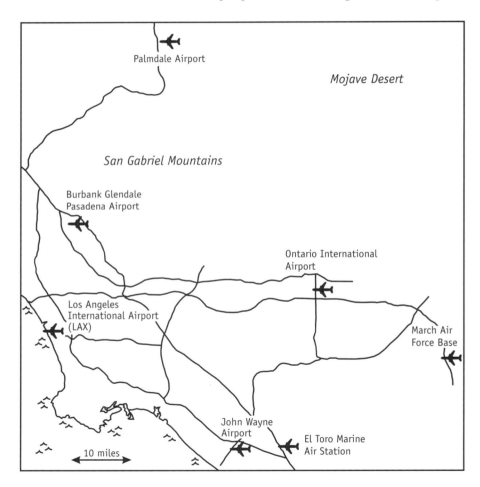

Figure 23.1 Major airports in the Los Angeles region.

International Airport is not the nation's largest single public works project being planned, it is close to it.

There is almost no argument that the United States could anticipate major growth in both passengers and freight, both from internal growth and from the rest of the world. In spite of the sinking spell in the Asian economies, the west coast is particularly focused on the Pacific Basin, which is widely anticipated to become a nexus of vigorous commercial and tourist activities into the long-range future. In spite of a bit of quibbling over precise numbers, estimates like doubling of passengers in 15 or 20 years for the Metroplex as a whole are widely accepted, and many expect even greater growth for freight.

The argument is less about the reality of growth than it is about how much of that growth has to be accommodated at LAX. Or, for that matter, how much of the growth might be lost if airport capacity falls short; there is deep concern that other cities would be happy to accommodate it. With modern commerce depending more and more on air transportation, many people view a potential failure of the Los Angeles area to capture a large portion of this Pacific Basin growth as a real threat to the general economic well-being of the region [7].

The folks who live close to LAX aren't against growth, they just want it at some other airport.

The Southern California Association of Governments estimated how the passenger traffic in 2020 might rationally be spread around the region [8]. They assumed that not only would all the existing airports absorb as many more flight operations as they appeared capable, they also assumed that the El Toro Marine Air Station would be converted to a commercial airport, becoming the Number 2 airport of the region. El Toro was assumed to capture 22 million passengers by 2020 (36% of the 1998 level at LAX). This would still be only roughly 15% of the total passenger traffic in the region.

Even with this major new airport, and with maximum feasible expansion of the other airports in the region, capacity at LAX still has to grow by some 75% if the forecasted doubling of passengers and freight into the region is to be accommodated.

El Toro is roughly 45 miles by helicopter south and east of LAX and 9 miles beyond John Wayne. Converting El Toro seems on the face of it to make a lot of sense. It has a third more acreage available than LAX and is located in the very rapidly growing Orange County, well positioned to serve that area's air traffic needs.

But many of the local population around El Toro were somewhat less than enthusiastic about an airport in their backyards. "Vehemently and implacably opposed" are the words that pop to mind. In fact, they didn't like it even as a Marine Air Station, which it had become years before when the county was only orange trees and bean fields.

One has to have a strong vested interest in an airport to willingly accept it as a neighbor.

Our story doesn't stop here: there are two other airports in contention for a significant role in the region. One is March Air Force Base at the southeast edge of Riverside, some 70 or so miles east of LAX. The problem with March is largely its distance from the major markets of the day. Given continued growth of the Inland Empire, however, much of which is industrial and a potential freight market, March is actually not badly positioned for the future.

The second contender is at Palmdale, which is a little over 50 miles north and east of LAX by air but closer to 70 miles by road. Back in 1969 the City of Los Angeles had purchased some 17,000 acres (wow!—LAX is on 3,500 acres and Ontario on 1,500) of flat desert land with the idea of making tiny Palmdale the home of a major international airport, connected by high-speed rail into the LA basin [9]. The large acreage and the thin population of the area was seen as the solution to the Not In My BackYard (NIMBY) problem and the general aversion to having large aircraft in large numbers landing and taking off over a well-populated city.

The plan for Palmdale never came together, at least partially because the transportation connection was so daunting to construct, given that Palmdale was separated from Los Angeles by the San Gabriel Mountains. And there was valid concern that the trip time—well over an hour with the then-available technology—would be too much of a pill for most travelers to swallow. There were also hints of hanky-panky with the land purchases that didn't help the cause.

But now in the late 1990s Palmdale is still favored by some rather influential people. And by now Palmdale is no longer so tiny, and one of the prongs of LA's growth has moved in its direction. The acreage available for the airport—almost 5 times greater than that of LAX—would allow substantial flexibility in layout. While NIMBY problems should never be underestimated, the very substantial acreage available around the airport to act as a buffer is a major, major plus for that location.

The dilemma of 2000

Thus our intrepid Planning Commission and the citizenry and polity of the LA Metroplex seems to be facing three alternatives:

1. Go ahead with a very major expansion of LAX;

2. Encourage maximum use of other airports in the region, including developing one or more new large airports, to accommodate much of the anticipated growth in air traffic AND also go ahead with a more modest expansion of LAX;

3. See much of the growth go elsewhere.

The people advocating the full-blown expansion of LAX largely base their case on the fact that LAX is already the dominant airport, serving as the primary hub for passenger traffic, and already has substantial facilities for the handling and transshipping of air freight. Since more than half the freight entering and leaving LAX travels in the bellies of passenger aircraft [10], both of these roles were seen to be much better served if both freight and passenger traffic were concentrated in one airport. The alternative is to force some passengers and some freight to travel from one airport to another to make connections.

While the hub function at LAX is not as big a part of its traffic as at some other of the nation's major airports, it is still significant—roughly a third of the passengers into LAX are there to catch another flight, nearly all of them on the same airline they came in on.

The people who object to the further major expansion of LAX base their case primarily on its impact on livability in the area. They cite having to cope with increasing traffic congestion on the road system and the environmental insult to neighboring cities.

They also raise the question of what to do next when sometime in the distant future LAX saturates again and we have no other alternative ways in place to absorb the growth. Will we have to bulldoze part of

Westchester or El Segundo to double LAX capacity again? And still far-ther into the future the same decision might be waiting to be made again. Even if all the answers aren't completely in hand in 2000, they feel it is not too soon to start thinking about alternative development paths rather than just facing the same what-to-do-now dilemma recurring forever.

The issues of how hubbing might be handled in a multiairport system and the losses in patronage if it is handled poorly are not the subjects of much public dialog at this writing, nor are the implications of the evolving air system route structure on that problem.

The issue is not a trivial one. One does not just wave a magic wand to distribute air traffic among airports the way one might want it; it is not just a question of deciding that $X\%$ of the aircraft are going to go here and $Y\%$ there. That decision is made individually by the various airlines, both those that already serve the region and those that might enter that market if it appears profitable.

A multiairport system without a dominant hub presents a kind of Gordian knot calculus to both the airlines, the aircraft suppliers, and, most important to us, the airport system designers. Alexander is reputed to have handled the original knot with a sharp sword. Now sitting at the end of the twentieth century most people do not see that there might also be a sharp sword in their future. We go on with our story.

The hub-and-spoke route structure

The hub-and-spoke route structure started long before deregulation. And contrary to many people's perception, the percent of passengers who have to change planes to reach their destination has changed very little with deregulation—for the nation as a whole up from about 28% in 1978 to nearer 32% in 1993 [11].

What has changed is its character. Prior to deregulation, roughly half of the passengers changing planes changed to another airline. Since deregulation, the airlines use hubbing as a competitive tool, with the various majors selecting certain airports to co-opt as their hub, and sched-uling their flights in and out for convenient transfer within their own air-line. If one airline can dominate a hub, it has an excellent chance of making sure that all the spokes feed its own big trunk route aircraft and not those of some rival airline. The ploy has been successful; now nearly all transfers are within the same airline.

When we talk about direct service and hub-and-spoke service, it might give the impression that we can sort out the two by sorting aircraft—paint one black and the other white. The actual picture is more subtle; all aircraft are shades of gray because it's not the aircraft that are hubbing it's the passengers (or the package or container): some are and some aren't.

We conjecture that every commercial flight that comes into LA has on board both passengers for whom LA is their destination and passengers who are coming to LA just to catch another flight—who are using LA as a hub. We also conjecture that while there may be a few passengers coming into one of the smaller airports that intend to catch another flight out, the majority of hubbing passengers will be on those planes going into LAX, simply because LAX has the longest menu of flights out.

From the airlines perspective other things equal, they would prefer to fly into the largest airport because that is how they can satisfy the most hubbing passengers (and from whom they collect an approximate 5% premium built into their ticket prices [12]).

There are, of course, other considerations that may well override the desire to cater to the hubbing passengers. Landing fees are likely to be lower at the outliers and air and ground traffic delays less onerous. Or there may be an overwhelming number of passengers who want to go to Orange County or the Inland Empire or someday maybe to Palmdale. Or the competitive picture may favor some airlines focusing on the smaller airports. Or maybe they just can't get into LAX.

The continuing evolution

Growing per capita income around the world is resulting in a growing population that flies, so more and more city-pairs can generate the traffic necessary to support direct service. And because there are more habitual flyers, there is a growing pressure for better service, including more frequent nonstop service rather than having to transfer from one aircraft to another. And again because of rising incomes, more and more people will be willing and able to pay for such service.

An airline's decision to respond to this changing market by offering more direct service rests primarily on four factors: (1) the number of passengers they estimate will fly that segment for the level of service that might be offered; (2) the compatibility of the segment with the airline's already existing route structure; (3) the likelihood of competing airlines also entering that market and diluting any chance for profitability; and (4) the appropriateness of the aircraft available for serving that route.

We want to emphasize this fourth variable. The threshold number of passengers needed to support profitable service depends both on what people are willing to pay and the economics of the aircraft available for that route. In a general sense, the whole route structure, molded by the sum of the individual decisions of all the commercial carriers, depends significantly on the characteristics of the aircraft the available technology can produce.

For example, we just don't know how to build aircraft that can carry five passengers per trip at a price very many people would be willing to pay, so we almost never see daily air service where there would be only very thin streams of passengers. (This kind of service will probably have to wait until we can operate without crews—that may be a little while yet.) Today very few "commuter" aircraft have less than 19 seats, and increasingly they have capacities from 29 to 70 seats. So passenger traffic streams of less than 15 to 20 daily passengers are unlikely to get any kind of regular air service.

The threshold for a new direct route segment depends on the distance to be flown. The ridership threshold for such a new direct route across the Pacific is not 15 or 20 daily passengers, it's more like several hundred passengers per day, simply because the only aircraft that have the necessary range capabilities are large aircraft.

As technology improves and seat-mile costs come down, the threshold required of the other factors lowers and more direct service is made possible. This is just one more situation where lowering the cost of small batch movements can improve service by making it possible to offer direct connections on thinner routes and more frequent service on well-traveled routes.

The airlines long ago recognized the growing desire for more direct service. The story of the Boeing's attempts to sell a super-747 was a clue. The 747 is the largest passenger airplane in the world and serves as a connector between the larger airports of the world. If continued growth in hub-and-spoke routing were anticipated, even larger aircraft to accommodate the even larger streams of passengers between such hubs would make sense. So Boeing had worked hard to satisfy that potential need with various super-747 designs. But finally, in early 1997, they threw in the towel: the airlines didn't want bigger airplanes [13]. The new airplane they had really wanted was the 777, smaller, not larger than the 747.

There may still be larger aircraft in our future, but they will probably be more a sign of growing traffic rather than a reversion away from trend toward more direct service.

We see this trend toward more point-to-point direct, nonstop service continuing indefinitely into the future for both passengers and freight and being accelerated both by growing incomes and by new aircraft with better small batch economics.

A zero-sum game?

If more direct service between city-pairs implies a significant decrease in the number of passengers who must hub to reach their ultimate destinations, then it has real significance for our problem of how to handle multiple airports serving the Los Angeles Metroplex, or any other large urban complex. If that were, in fact, the case, there is no particular advantage of being the largest airport with therefore the longest menu of destinations from which the hubbing passenger can choose. The choice among airports would depend entirely on the draw of the area it serves; one picks the airport most convenient to their final destination.

So does the growth in direct service portend the ultimate disappearance of hub-and-spoke routing? The answer, we think, is a resounding NO; it is not a zero-sum game.

Think of the admittedly somewhat artificial situation in which all cities and towns were "growing," growing in the sense of having more and more people who fly and more and more need for air freight service.

With growth in potential ridership, new origin-destination pairs that can support direct service come into being all up and down the spectrum of city sizes. The introduction of direct service among these city-pairs obviously reduces the number of passengers that have to take indirect routes to reach their destinations—who have to hub.

But this growth is also increasing the number of passengers using indirect routes; it is not confined to just those origin-destination pairs that become eligible for direct service; it is also happening among those pairs that have not yet reached the eligibility threshold with the available aircraft. Thus, as the number of direct flights increases, so does the number of passengers (or amount of freight) using indirect routes.

Second, as smaller towns proliferate and grow, or as small aircraft economics improve, more and more service points will be added to the air system, all depending on hubbing for access to the rest of the world.

Now we have no conviction about whether people and freight flying direct nonstop will grow faster than people and freight still flying hub-and-spoke, or from points being added to the system from the bottom. With more direct service, we will get more airports acting as minihubs, convenient connection points between city-pairs not quite ready for non-stop service. But if the air system continues to grow as expected, both the top end and the bottom end will grow.

But either way, the trend is likely in the long run to lower the average size of aircraft in the total commercial fleet. As this happens, the number of landings and takeoffs will grow faster than the numbers of passengers flying.

This line of reasoning leads us to believe that the growth in direct service will not eliminate the question of how to provide convenient and efficient hub service in a metropolitan airport system as the traffic is increasingly distributed in multiple airports. We still have the problem of how to cope with hubbing passengers and freight if we spread flights among separate airports.

The Airport Planning Commission of 2020—and what actually happened in the prior 20 years

Hindsight is wonderful!

We suggest at this point that the reader resign his or her membership in the Airport Planning Commission of 2000 and sign on for our new

Planning Commission of 2020. This gives a much better view of how the folks back at the turn of the century sorted things out.

Just because its 2020, we are not so old that we can't remember flying from the west coast to Washington or Boston or New York by changing planes in Dallas–Ft. Worth or Chicago or St. Louis or changing in Miami to go to South America or southern Europe.

From those memories we devised a law just as immutable as the *"The American shopper will not walk more than 600 ft from his or her car"* that we cited in our minicity chapter. Our law of hubbing is similar: *"No matter what gate your plane arrives at, your connecting flight is always at the opposite end of another part of the terminal."*

As we recall, the high-pulse-rate time—carrying one bag and a briefcase—from one end of United Concourse A in O'Hare to the other end of Concourse B averaged 12 min with a standard deviation of about 3 min. We are reluctant to quote the longer, normal-pulse time because we experienced it too seldom to have arrived at a statistically valid figure.

While the gate-to-gate time at the outlier terminal at Dulles was only 5 to 10 min, throw in the bus trip from the Main terminal and 20 to 25 min is not a bad figure. Dallas–Ft. Worth was usually a bit less, more like O'Hare in our memory.

Omitting the time allowed to compensate for the uncertainty in flight arrival, the price in terms of time paid for hubbing was generally 20 to 30 min at nearly all airports large enough to be a really desirable hub. Sometimes we were luckier, but we couldn't depend on it.

Improving technology and the sharper sword

Ah! If the folks in 2000 facing the dilemma of what to do could have put Palmdale only 20 min from LAX, then it really wouldn't matter where the flights came in. A plane could land at Palmdale, and the passengers wanting to fly out of LAX could be there in 20 min. Of course, 15 min would be better, but even 30 min might be acceptable without too much complaint. This would take some of the long-run pressure off the continuing need for a major expansion of LAX since new growth could be accommodated at Palmdale with only minor penalty to those passengers that needed LAX access or vice versa.

The obvious answer, and not a new thought at all, was tying the airports together with high-speed ground transportation. High-speed rail as talked about, but without enthusiasm. We would have been dubious of the suitability of this answer anyway; we would want to go very slowly in trying to squeeze the last drop out of a nearly 200-year-old technology when a very promising alternative is already well over the horizon.

To our knowledge, very few people in the late-1990s were thinking of magnetically levitated and propelled vehicles (maglev) as a serious option in LA. After a quarter century of research, by 2000 both Germany and Japan had 300+-mph vehicles running on test tracks and plans for operational systems within the next decade. Germany specifically had plans to open a 300-mph maglev system for operational use between Hamburg to Berlin in 2005 [12]. Somewhat behind these two in development was the American system already mentioned in Chapter 19. Being behind is not all bad: it offers the advantage of learning from what has gone before, and less-than-optimal decisions from earlier thinking are not already built in.

The fact is, though, that the maglev option was really not available in 2000. At that time an operational maglev system was still a little too risky to commit to. There was a while to wait.

There were other considerations that required a little faith in what the future might bring. It was generally accepted that the FAA would finally succeed in up-dating its equipment to at least late-twentieth century standards. This thought largely alleviated concerns that more air traffic into multiple airports would exceed the ability of the air traffic system to handle safely. It was widely expected that approach and climb-out paths could be made much more flexible, and the prospects for tighter spacing of aircraft to increase runway capacities a bit more seemed good.

Over the longer term there was also the likelihood of more environmentally friendly aircraft. The gains from the 707 to the 777 in noise reduction, emissions, and in flexibility in climb-out and approach paths were remarkable, and there was little reason to doubt that there were still further improvements to come. But the process of renewing the fleets was recognized correctly to be slow, so the promise of more neighborly airports was for the longer range future and had little impact on the attitudes of the day.

The upshot: what actually happened

There was much pulling and tugging between the LAX expansion advocates and the citizenry of the neighboring cities. In the end, it was generally recognized that the case for some expansion of LAX, at least in the near-term, was too compelling to not begin the process. It would take time to bring other airports into being, and the growth wouldn't wait. Failure to accommodate this growth in turn would have a negative impact on a region's ability to compete as a major commercial gateway to the Pacific. The ultimate cost would be a loss in commercial momentum and a slowed growth of regional employment.

But the environmental and quality-of-life concerns of nearby cities were real. One sore point was moving the south runway nearer to El Segundo. In the end a compromise was struck in which LAX backed off some on some of their expansion features but did add a new commuter runway, thus relieving most of the commuter traffic from the main runways and thereby expanding airside capacity. Together with the additional capacity that came with improved Air Traffic Control, this was enough to meet the immediate needs of the region.

Never underestimate our ability to just muddle through.

It was decided that a major airport at Palmdale really was, in the long run, going to have to be included as a major element of the Metroplex Airport System. From the beginning the design of the physical layout of the airport was based on the notion that it would be in an almost continuous state of expansion as its role in the region's complex expanded over the years.

The initial implementation was for those elements of the design necessary to make Palmdale a major hub for dedicated freight aircraft. Remember, though, that much of the freight coming into the region was coming in the bellies of passenger aircraft, which perforce were mostly going into LAX. So it wasn't in the cards to make Palmdale the freight airport, only a supplementary airport, but with prospects for a larger and larger role over time as freight traffic grew.

In the beginning transportation to the LA basin was to be handled by truck, but with recognition that ultimately the airport would be served by some form of high-speed, nonhighway system. The usage of State Highway 14, the primary Palmdale–LA connector, was already nearing its capacity, so an expansion of that route as well as some of the freeways within the basin were included in the overall project. It was not cheap.

The conversion of El Toro to a commercial airport was a tougher decision. The opposition on environmental grounds was fierce, but it became increasingly apparent that John Wayne, which was already expanded to its practical limit, would not meet the needs of the growing county. In the end, the conversion to a commercial airport was begun.

Like Palmdale, the ground transportation needs of El Toro were also met initially by the road and freeway system but with recognition that some higher speed connection to LAX would be needed before the greater efficiency and user-attractiveness of an integrated, regional airport system could be realized. Provision for this additional transportation element was included in the initial designs.

Then in 2007 the Berlin–Hamburg maglev line opened, with its transrapid system providing 300-mph service (cruise speed) with magnetically levitated and propelled vehicles. At the same time the American version of maglev was getting close to becoming operational in the first leg of what was to become a 55-mile line linking the Orlando International Airport to the Kennedy Space Center and the Canaveral Cruise Ship Terminal. It was being built to carry both passengers and freight at top speeds of 300 mph.

Los Angeles officials, engineers, and other technical people had already made numerous inspection trips to study these new systems and had begun serious evaluation and deployment studies for connecting LAX with both Palmdale and El Toro. By 2008 plans were in place to begin construction of a new maglev system connecting LAX and these new airports.

With the prospect of much faster connections back into Los Angeles, more emphasis was placed on making Palmdale an important airport for passengers as well as freight. A new air system access terminal (no runways) was added to the high-speed line in Van Nuys to better serve the San Fernando and Simi Valleys.

The line was elevated most of the way, generally along the 405 Freeway and State Highway 14, but leaving it in places to minimize curves—which were a real problem on 14 as it went through the mountains. An overhead structure was used that could run down the existing highways but could leave it where appropriate. It vaguely resembled a double decked freeway but was much less massive. The vehicles—"cars"—were considerably lighter than equivalent-capacity high-speed rail.

The cars were designed to operate automatically, normally 4 min apart, but capable of every-minute service. The capacity of the line could be altered to match demand by entraining or dropping vehicles. Some cars and trains were designed for freight.

Express service was available. The possibility was considered of achieving this by having trains form and disassociate in motion, so that, for example, everyone going from Palmdale to Van Nuys could get in the Van Nuys car, which would separate from the other cars and drop off-line at that terminal. It was decided to plan that capability for the future when systems for automatic baggage sorting and handling had been fully developed and begin with the simpler approach of having different cars or trains designated for different destinations at departure.

Operational cruise speeds of 250 mph were planned initially, resulting in portal-to-portal times from Palmdale to LAX of about 15 min for the express.

To be honest, this in itself would not make Palmdale and LAX 15 min apart for the average passenger. If nothing else changed, it would probably take a passenger 5 to 10 min to get to the maglev terminal and certainly at least that in LAX to get to the correct gate. And if the maglev vehicles left 4 min apart, there is another 2-min average wait for service. If nothing else changed, arrival-gate-at-Palmdale to departure-gate-at-LAX would be closer to 30 to 35 min, enough time to begin to be a bit onerous.

But the designers recognized the problem and planned track for each maglev vehicle to move through the terminals, one floor above the passenger gates, making short load-unload stops at convenient points that were accessed from the passenger gate level by an escalator and elevator. That vehicle, after "sweeping" the terminal for passengers and transferring luggage that had been checked through, would move onto the main maglev line and accelerate to cruise speed. This technique reduced the gate-to-gate time for Palmdale to LAX (and vice versa) to 20 to 25 min. Still a little worse than other large airports of the times, but a lot more interesting.

The El Toro connection had stops at John Wayne and at a new air system access terminal (no runways) in Long Beach.

The line ran generally along the 405 Freeway, but—analogous to the Palmdale leg—leaving it in places to minimize curves. The integration with the new El Toro terminals were also similar in approach to that used in Palmdale.

The process of putting together a funding package, using largely federal funds, had been begun before the final decision to go with maglev, so final approval came relatively quickly. Construction was begun on the system in 2010. The system was operational in early 2014.

By this time in 2020 plans for another maglev line connecting the El Toro line with Ontario and March were being completed. Construction of this new line was begun in 2016, and was completed last year.

It has been a busy 20 years.

The view from 2020 looking forward

Here, fellow members, we of the Airport Planning Commission of 2020 outline the situation looking forward.

Almost everyone expects major growth in both passengers and freight in the future. An estimate like doubling both by 2035 or 2040 is widely accepted, and there are many who argue that it could easily happen sooner, that we are underestimating the impact on air traffic of the dramatic improvements in communications that are still coming into being and the more vigorous market economies in South America and the Pacific Basin.

Plans are nearly complete to close down the south runway in LAX and depend more on the runway capacity at Palmdale and El Toro. The idea is to concomitantly relocate LA's Hyperion Sewage Treatment Plant to a location east of Sepulveda. (The Hyperion is on the coast on the south side the airport.) These two actions would restore a major section of attractive beach-front property for other uses.

The techniques to produce more electricity from sources that are not fossil-fuel dependent seem to be getting closer, maybe in the next decade. One of the real objections when maglev was first introduced was that the very high speed system used electricity prodigally. If abundant cheap power actually becomes a reality, then this objection to still higher speeds will fade. Studies are underway to evaluate the feasibility and cost-effectiveness of increasing the cruise speed of the maglev system from its original 250 to 350 mph and to expand its capacity. These studies are expected to be available by 2026.

The view from 2040?

Trying to foresee the 20-year plan for 2060 is really a bit beyond our job description. But maybe we can make a few guesses.

Back in 2035 there was a citizens, group from the cities near LAX that was raising the question of why not close down all LAX runways and convert the airport into an access terminal only. This would restore the whole coastline all the way to Marina del Rey to recreational use. They envisioned putting all the air operations at Palmdale.

While this looked totally impractical from our view in 2020, we can't rule out the possibility that by 2040—40 years into the twenty-first century—the possibility might be real. The same kind of talk went on from time to time about closing the runways at John Wayne and El Toro, diverting that air traffic to Ontario and March, or to a big new airport at Camp Pendleton between Oceanside and San Onofre. Right now it's just wishful thinking; interest will die down in time.

It has become very clear that the shift made back in the first decade of the new millennium from the notion of just juggling flights among the different airports in the region to the idea of a regional airport complex integrated through a very high speed ground system has opened up a whole new vista of flexibility in airport planning, resulting in an airport system that can much more easily adapt to the changing needs and constraints of the evolving Los Angeles Metroplex.

It is also gratifying that the same approach is well underway in lots of other cities and city-combinations as well.

References

[1] Radcliff, J., "Alternate Runways," Torrance, CA: *The Daily Breeze*, April 14, 1997.

[2] Los Angeles World Airports, *International Airport Statistic–Volume of Air Traffic*, compiled from Airline Traffic Reports, U.S. Customs and the FAA, 1999.

[3] Los Angeles World Airports, *General Information*, Los Angeles, 1999, www.lawa.org.

[4] Newton, J. "3 Huge Projects Could Define 21st Century L. A.," *Los Angeles Times*, December 14, 1997, p. A44.

[5] Grad, S., and L. Munoz, "Southland Airport Planners Face Frustrating Paradox," *Los Angeles Times*, March 1, 1998, p. A3.

[6] Newton, J. "3 Huge Projects Could Define 21st Century L. A.," *Los Angeles Times*, December 14, 1997, p. A44.

[7] Erie, S. P., "Developing Outlying Airports Won't End Region's Crisis," *Los Angeles Times*, May 10, 1998, p. M6.

[8] Newton, J., J. Wilgoren, and L. Munoz, "Battle Over LAX Expansion Leaves Behind Turbulence," *Los Angelers Times*, May 29, 1998, p. A26.

[9] Newton, J., J. Wilgoren, and L. Munoz, "Battle Over LAX Expansion Leaves Behind Turbulence," *Los Angelers Times*, May 29, 1998, p. A26.

[10] Los Angeles World Airports, *General Information*, Los Angeles, 1999, www.lawa.org.

[11] Morrison, S. A., and C. Winston, *The Evolution of the Airline Industry*, Washington, DC: The Brookings Institution Press, 1995, p. 22.

[12] Morrison, S. A., and C. Winston, *The Evolution of the Airline Industry*, Washington, DC: The Brookings Institution Press, 1995, p. 48.

[13] Associated Press, "Boeing Will Stop Work on 'Superjumbos,'" *Los Angelers Times*, January 21, 1997, p. D1.

[14] Geerlings, H., "The Rise and Fall of New Technologies: Maglev as Technological Substitution," *Transportation Planning and Technology*, Vol. 21, 1998, p. 272.

Part VI

The New Millennium

24

Communication and Transportation

Transportation and communications are certainly close cousins. They support the interactivity of people, places, and activities; and their services are complementary and competitive. What are implications of the rapid improvements in communications services?

It's tempting to start looking for analogies between transportation networks and communication networks or to play with words by noting that communication is just the transport of information. We toyed with those ideas a bit and concluded that it wasn't getting us very far. Clearly there are big gray areas: transporting a letter is communication, when Uncle Robert comes to visit us from China he brings us news, and when the train doesn't reach Deadgulch on time it may be telling us that the buffalo are blocking the track again. But this contributes little to our understanding.

The real drama is not about gray areas and analogies. We are clearly in the middle of a revolution in communication and computational technology, a revolution that many believe could be as important to humankind as any prior metamorphosis the world has known. Whether we consider this Act 2 of the Industrial Revolution or an entirely new play really doesn't matter. What does matter is the way it might impact how the world lives, works, and interacts.

On the technological side, Act 1 was largely driven by what we now call thermodynamic and mechanical engineering, the generation and harnessing of mechanical power derived from heat. An important supporting role was played by the telegraph, the telephone, and the postal system; but center stage was occupied by the new and dramatically better forms of power, by concomitant advances in transportation, and by increased mechanization of productive processes.

The trigger this time is an expanded understanding of solid-state physics, which, through the technologies that exploit its principles, has after many years of incubation given completely new meaning to computation, communication, and sensing. It pains us slightly that in Act 2 these are the developments that are playing the lead, and transportation, while also exploiting them in very important ways, is now a supporting actor that is absolutely crucial to the plot but not on center stage.

What's going on?

The magic word in communication is bandwidth. Until a few years ago, communication largely meant the telephone system, transmitting a thin stream of data—the human voice—over complex circuits for the period of the conversation. Increasing bandwidth is the technical way of saying that we can transmit larger and larger quantities of data faster and faster—as big batches organized in streams of short bursts—and, fortunately, at lower and lower costs.

Computers around the world can now be usefully networked, so communication is not limited to people to people or TV station to TV. An Automatic Teller Machine in Spain can effectively take money out of my bank in Los Angeles (hopefully only when I want it to). Direct communication among people is no longer just voice; a surgeon in New York can watch a heart transplant in Cairo in real time, receiving all pertinent information, and consult on the on-going procedure. Such high data traffic was once small compared to voice traffic, but this relationship is very radically reversing.

This truly remarkable leap in communication capability has come about as the result of a succession of fundamental technical advances such as the transition from analog transmission (varying the waveform of the carrier signal in proportion to the sounds of the voice) to digital wherein all signals are translated into codes of ones and zeros; continually improving the computational capabilities that permit much more sophisticated processing of this digital stream; improving the ability to exploit higher electromagnetic frequencies; and the placement of satellites to cope with a round world. These developments not only let us use very broadband transmission conduits like fiber optic cable but they open the door to increasingly versatile wireless. We're even learning how to push more data over copper wires than we've ever been able to in the past.

One way that costs are being lowered is through the use of wireless transmission as a substitute for wire in the "last mile" into the home or business. This "Wireless Local Loop" or "Local Multipoint Distribution" by lowering the cost barriers of providing service makes it much easier to bring better communication to the vast underdeveloped or less densely populated portions of the world such as China, India, Africa, and South America.

There are some inconveniences in living on a round planet that has mountains and valleys and buildings on its surface; specifically, the high-frequency wireless transmissions that are necessary for high bandwidths don't bend, don't bounce, don't go through solids, and are eaten up by moisture in the air. Low-Earth-orbit satellites (LEO) some 400 miles above the Earth provide the clear transmission paths that accommodate these properties without the almost half second time delays associated with the very long trip out and back to the geostationary satellites, 23,000 miles up. The speed of light is fast, but 46,000 miles is a long trip, and this delay is an eternity for transmissions.

So we can summarize. The ability to move data in very large batches and the increased versatility of communication that results are the first dimension of change in our brave new world. The second is the dramatic and ongoing drop in the cost of such communication. The third dimension is ubiquity, worldwide access to this information flow.

It takes two to tango

How does transportation fit into this picture? We have already brought out throughout this book that the rather remarkable new technologies that are driving the computation-communication revolution are also enabling new capabilities in the transportation system itself. These are reflected in our visions of automation; in our visions of better traffic control on our roads and railroads and in the air system; and in improved operational control of the organizations that foster and operate our transportation systems and companies.

We see more efficient designs in vehicles themselves, designs that take advantage of this explosion in capability to automate the hundreds of individual functions that define their total performance, freeing them from almost total dependence on mechanical engineering skills of years gone by.

The primary interest to us here is the interplay between communication and transportation. In the popular press this is most often thought of in terms of one substituting for the other, especially substituting communications for commuting or telecommuting. Teleshopping and teleconferencing are also mentioned. Communication has become a technological fix for congestion and lots of other things.

But it seems to us that the more important and interesting question involves the synergies, their combined impacts on all those economic and social processes in which they both are essential and integral elements. As Salomon stresses, there is the potential for interactive social and economic change [1].

Synergies and impacts: looking around and looking back

One might think of communication as a creator or modifier of the motivation or incentive for transportation. It is, of course, not the only one:

communication doesn't create the demand for petroleum, but it does influence where we go to get it and therefore how we transport it.

Better and more complete communication and data transfer should certainly be able to increasingly substitute for some kinds of travel. The routine conference no longer demands physical presence, and presentations complete with voice, video, and supporting data can be made from anywhere on the globe. For knowledge workers telecommuting becomes an option. Clearly, this substitution for face-to-face interaction reduces the motivation for travel [2].

At the same time, improved communications vastly improve opportunities for individuals and institutions to work together. Nothing new here, Carolyn Marvin reported that over 100 years ago a Judge Taylor remarked that the telephone had introduced the neighborhood without propinquity [3]. One's neighbors are the folk one phones, and they may or may not be next door. Anticipating Judge Taylor, Michael Shnayerson reported that Prince Albert remarked at the opening of the London Exhibition in 1851, "... communications and transport have erased the vast distances that once separated mankind [4]."

And also at the same time, better and richer communication capabilities coupled with increased opportunities stimulate entrepreneurial travel. At the heart of entrepreneurship is the search for and creation of new markets or new coalitions. These usually portend new human relationships, and the building of these relationships is likely to continue to require face-to-face interaction. We suspect that improving the reach and quality of communication will stimulate this entrepreneurial travel. Again, nothing new here. Incorrectly anticipating that travel would be decreased, early U.S. railroads tried to block the spread of telegraph services until the Baltimore and Ohio Railroad allowed the use of its Baltimore Washington right-of-way for telegraph wires [5].

So communication doesn't automatically keep people at their desks, it might also lure them away. We are told about the companies that have started to operate in the "hot-desk" mode: executives that are on the road so much that leaving empty offices unused is wasteful, so they time-share offices. Their laptops are their filing cabinets and their fax machine and Internet access. Obviously, for these key people in increasingly geographically dispersed businesses, having access to very good communication has not cut their demand for travel.

And then there is pleasure travel. We just can't bring ourselves to accept the extreme science fiction vision of people living out their lives in a virtual world, experiencing the Seychelles or even Philadelphia from an armchair. There are too many dimensions of lives and vacations that require reality. We just don't believe that folks will settle for a digital stream as a substitute, no matter how clever.

We can only guess whether the decrease in routine travel will be more important than this growth in entrepreneurial and pleasure travel. Our past experience tells us that on balance better communication is a net stimulus, but we have to admit that our past experience probably does not prepare us for the rather remarkable communication and media capabilities forecast for the future.

Globalization

We characterize the communication-transportation impact with a simple two-step process. In step 1 cheaper, faster, better, and more ubiquitous communication makes people aware of situations and opportunities that were out-of-sight in a less communicative world. Step 2 does something about it, and human nature being what it is, that will often mean going there to make it real, perhaps just to establish that bond on which so much commerce depends, or to take the more tangible action of setting up a new import or export channel, siting a new production facility, or exploiting new sources for raw materials, for example. When these involve physical movements, then an incentive for increased goods movement is generated. Most of these actions fall under the rubric of globalization because transportation and communications advances have extended to a world stage.

Globalization—the growing economic integration of nations around the world—is the most obvious of the trends identified with the communication revolution. It is usually viewed as a new phenomenon by the history-impaired, but it in fact is just the latest label on a process that started when Village X traded some of their clay pots for Village Y's bone awls. To be fair, globalization implies much more than just an expansion of trade; there are dimensions of integration that are, in fact, new in the sense that their earlier manifestations were on too small a scale to matter much.

Until the last century or so, the opportunity for economic integration derived entirely from the available transportation because communication independent of transportation did not exist. Now an independent and versatile communication is a precursor, serving to establish motivation for transportation, which then serves the role of enabler.

The actual picture is complicated. While communication, transportation, and Adam Smith's invisible hand can motivate and enable interactions between geographically separated centers, governments sometimes exhibit the iron glove by erecting barriers to this interaction. So commercial integration of the nations of the world has not increased either uniformly or along a smooth trajectory. We conjecture that the increasing dialogue between nations that has been enabled by better communication and the increased trade that it has motivated are likely over time to reduce these regulatory barriers.

We have also seen political integration follow along behind greater economic integration, again not smoothly and not consistently—and not always happily. But city-states have now largely become nations, and small nations have consolidated into larger ones. The economic integration of the United States is essentially complete. State boundaries remain important for other reasons, of course.

Perhaps this describes the situation. Until the last decade or so, transportation and communications have expanded in ways that were partly complementary and partly substitution in character. They have marched in lockstep as the developments they have seeded/enabled have asked for their services. The two-step dance has had several dancers, so to speak. Data on the growth of passenger transport and communications in France support that observation [6], as does information from Sweden [7]. In both cases, as elsewhere, communications and transportation have grown in the same fashion.

Now we find ourselves at the leading edge of the twenty-first century with communication capabilities that exceed our imaginations of a few decades ago fueling and accelerating the globalization trend.

It follows that the task for transportation has a keep-up character. This is our conjecture, of course. And transportation will keep-up because of the way it and communications work together. The question is how well.

One manifestation of today's developments that is already fairly clear is what we might call the distributed factory. In the distributed factory the work in process is moved from place to place during the production process to take advantage of special capabilities or lower prices for each of the individual steps in that production. Better communication will contribute to this acceleration by facilitating the coordination of such processes, and better transportation will enable it.

Lowering the cost, particularly for the movement of small batches, obviously reduces the cost barrier to exploiting the advantages of this "distributed production." Increasing the speed and frequency of service (the other side of the small batch coin), the reliability, and the predictability of service lowers total costs by reducing the work in process inventory—the time lost as expensive parts are moved from one process location to another or held in local inventory to guard against failures in transportation. The combination of better communication and better transportation opens much broader vistas, and probably some completely new ones, for optimizing production processes.

The ability to transport goods quickly and cheaply to exploit local advantages in the various steps in production processes has the side effect of trending us toward a worldwide equalization of labor costs, corrected for local disparities in productivity.

It seems to be reasonable to think that analogous improvements in communication might have the same impact on the world's knowledge workers, because it widens the market into which each can sell their expertise. Rather than such workers being constrained to sell their talent into the local market, we will move toward a world market where all kinds of knowledge skills are accessible to all. A small example is in India, where software is written for American organizations.

The downside

Adam Smith brought into clear focus the productivity advantages of the specialization of labor. But this specialization created a new need, specifically, to coordinate the individual activities that make up the whole.

On the scale of activity with which Mr. Smith was acquainted, this posed no particular problem and was probably accomplished with a little conversation. We have dramatically outgrown that scale today. The increasing specialization of functions and activities—and the increasing scale of these functions—has made this facilitation and coordination of the interactions among them a crucially important part of the workings of our world.

The computer, frequently in concert with better communication and transportation, has revolutionized this coordination function. In many cases it has automated it, taking the human out of the loop, substituting computer control, enhanced by computer-to-computer communication. We see this happening in all kinds of traffic control: highway, air, rail, and water. We see it in computer-controlled factories, in commerce between financial institutions, and in retail outlets. We see smart washing machines and smart toasters. It's a more tightly controlled, more productive, smarter world. It's a world undergoing a control revolution [8].

We read in science fiction about computers taking over. In the sense of substituting for human thought, we doubt it. In the sense of making us humans dependent on them in nearly all aspects of our lives, they have already. Computers are tools we have exploited to produce huge improvements in the way our world works.

The downside is obvious: we are very vulnerable to those occasions when they do not work, when we have to lapse back on the old ways to carry out functions that are now almost totally dependent on the new ones.

This is not a new circumstance in the history of the world. We become vulnerable to failure whenever a new technology renders an old one "obsolete." Few of us could make a fire with flint and steel (and many of us have a hard time with matches). But that is small comfort. We are vulnerable, we are forced to recognize it, and where economically feasible we put backups in place to ease that vulnerability. The road we travel, faster and easier and more satisfying than ever before, still has potholes, both metaphoric and real.

And we are vulnerable in another way. We have said coordination when we could have just as well have said control. We think it was just fine that the telegraph enabled controlling steel making by tying it to markets and controlling the movement of trains. But we are not so certain about the ways information systems and price setting sometimes worked out. The expression monopoly pricing rears its ugly head.

We are vulnerable to upsetting the delicate balance between those who govern and those who are governed.

Adolescence

We earlier made the points that the spatial organization of our world is constrained by the capabilities of our transportation and that for most of history that largely meant walking, sometimes running, and sailing ships. Transportation and communication—the same thing—was almost always at speeds under 3 to 4 mph. For thousands and thousands and thousands of years of human history our habits and our institutions were constrained and, to some degree, shaped by these simple facts of life.

Then in the last half of the nineteenth century the railroad, the telegraph, and, less dramatically, the steamship began loosening these ancient transportation constraints. By 1900 the telephone was beginning to reshape personal habit patterns and institutional organization. We began mass-producing cars in the first decade of the twentieth century; and in 1920s the United States went from one car for every 30 or so people to one car for every five (essentially one per family) and radio was becoming a staple. The first successful airlines appeared in the 1930s, TV in the 1950s, and in the 1960s man walked on the moon. The 1970s produced the first personal computers. We are now living in the middle of an explosion of new communication/computation capabilities and looking forward to concomitant improvements in our transportation capabilities.

We, our parents, and a few of our grandparents are the first human beings on the planet to live in a world in which the age-old transportation/communication constraints on the spatial organization of our infrastructure, our daily habits, and our institutions have so significantly dropped away. We have been given the means, the options, to do it differently than it has been done for millennia, but we have been given these options without a book of instructions.

References

[1] Salomon, Ilan, "Technological Change and Social Forecasting: The Case of Telecommuting as a Travel Substitute," *Transportation Research*, Vol. 5C, No. 1/2, 1998, pp. 17–45.

[2] Mokhtarian, P. L., "Telecommuting and Travel: State of the Practice, State of the Art," *Transportation*, Vol. 18, No. 4, 1991, pp. 319–342.

[3] Marvin, C., *When Old Technologies Were New: Thinking about Electric Communication in the Late Nineteenth Century*, New York: Oxford University Press, 1988, p. 66.

[4] Shnayerson, M., *The Car that Could: The Inside Story of GM's Revolutionary Electric Vehicle,* New York: Random House, 1996.

[5] Stover, J. F., *History of the Baltimore and Ohio Railroad*, West Lafayette, IA: Purdue University Press, 1987.

[6] Grubler, A., *The Rise and Fall of Infrastructures: Dynamics of Evolution and Technological Change in Transport*, Heidelberg: Physica-Verlag, 1990, p. 256.

[7] Batten, D. F., "The Future of Transport and Interface Communication: Debating the Scope for Substitution Growth," *Transportation for the Future*, D. F. Batten and R. Thord (eds.), New York: Springer-Verlag, 1989, pp. 67–86.

[8] Beniger, J. R., *The Control Revolution: Technological and Economic Origins of the Information Society*, Cambridge, MA: Harvard University Press, 1986.

25

Optimism

"It is not the strongest of the species that survive, nor the most intelligent, but the one most responsive to change."

Charles Darwin

Centuries ago maps showed a world as flat as a pizza box. Unexplored territory was marked by sea serpents lurking at the edges of the seas and ferocious, nightmarish animals on the land. The imaginations of sailors and navigators were constrained to the top of this box and they were terrified of the unknown. We still have our fears and our blind spots, still cling to the known, are nervous with the unknown, and sometimes tilt at windmills.

Thinking outside our figurative pizza box still remains difficult. But the once-fearsome animals have largely turned into economic and environmental worries. Will it cost too much? Will it make money? Will it damage the ecology, ruin the air, or make too much noise? Will it run too close to my house? All very legitimate questions, but no longer fearsome creatures with supernatural powers.

An important difference from that earlier world and today's is the notion of change, a notion that is only a couple of hundred years old. People not so long before Stephenson's time took the world as given and unchanging. Now change is a part of life.

We borrow from the introduction some of the words that express perhaps the major theme of the book. Nearly all the productive or social processes of working and living involve transportation, either as an integral part of the process itself or in the activities on their periphery. Better transportation enables improvement in almost all these processes, often dramatically and often in ways not now imagined. It lets us do old things in new ways or entirely new things.

So advancements in transportation can make our collective future better over time in ways that transcend the transportation itself. That is where the real payoffs are: what transportation can do to improve all the other aspects of our lives.

Our chapters drew forward visions for passenger and freight services in towns, the countryside, and in the air. From technological opportunities to generate cheaper, faster, better services for users, we have hypothesized some development paths—maps—to the future. We were guided by a sense of what society seems to want or need in the way of services and constrained by a sense of ecological, energy, and economic realities. The maps are not prescriptions. Far from that, they are possible opportunity-openers for folks who want new options for work and play.

We think it appropriate to recapitulate those points that stand out most in our minds about the directions of change and the lessons they might imply.

Perhaps the most important reality we have to accept is that the automobile is by far the dominant means of personal transportation in the United States and will continue to be as far as the eye can see. Thus the ambiance of our cities and the quality of our lives hinge in a very direct way on our ability to accommodate and civilize it. Our energies are far better spent in learning how to better live with it than in fighting it.

The automobile dominates because it fits into the inexorable pull for the personal: it lets us go where we want, when we want, with the radio tuned to our channel, with the company we want, and usually to go there faster than is possible with any other transportation available. Because of its superiority over competing modes of transportation and its widespread availability, the automobile has enabled the urban environment to reshape itself until for most of us it is now the only practicable personal transportation choice available. Tomorrow the pull for the personal may expand our desires—our choice of groceries delivered at home when we want them, our personal airplanes, and maybe our very own personal elevators. Perhaps these services are coming closer, but as far as we can see the automobile will still remain center stage.

We often sense a kind of throwing-up-the-hands, what-can-we-possibly-do attitude toward the problem it poses, the urban congestion problem in particular. As we have discussed at length, we think there is much that can be done.

Indeed, and as we have discussed, society has tempered the ways congestion affects everyday life and has worked to tame energy and emissions problems. Safety has improved. The fatal accident rate for motor vehicles decreased from 4.9 per 100 million miles of travel in 1970 to 1.71 in 1997 [1]. The total number of deaths has decreased by about one-third since 1960 in spite of the increased numbers of automobiles and the miles they travel. Similar patterns hold for all the modes. Here, as elsewhere, the future looks good.

Our Martian would tell us that in the evolutionary scheme of things, the automobile and the truck are becoming rather long in the tooth, so to speak: they are about a century old. But from our perspective they are just teenagers. They are still in a state of continual, incremental change, and there are readily foreseeable changes that may well go beyond "incremental": new forms of power; total automation; new roles, particularly in freight applications; and the prospect for major changes to the road and traffic management systems themselves. It's a highly inappropriate time to throw up our hands; it is time to think ahead, to take advantage of the inevitable continuing investment in the system to not just maintain and replicate but to improve along every dimension of performance that we know how. We have already offered some food for thought.

There is a distinct and positive role for transit as we know it, but we fear it is too often promoted and applied inappropriately. Transit was an outstanding success when it competed with walking or with horse-drawn conveyances at the turn of the last century, but is no match for the car at the turn of this one. Today transit only belongs in those situations where, for whatever reason, it is truly superior to the car. Wishing nor exhortation won't make it so where reality does not. We fear that too many efforts to introduce transit contrary to peoples' real preferences are, and will continue to be, expensive failures.

The automobile is too street-space-hungry to serve well in high-density environments, so we do need systems that are better tailored to those situations. Buses above ground and subways below ground serve that role today, but there is lots of room for new ideas and new systems. Personal rapid transit systems are certainly candidates, as are the new types of personal and freight vehicles that we have also mentioned. These or similar systems may in the future also open the door to entirely new spatial arrangements and living styles.

Along with the motor vehicle and the remarkable advances in computation, automation, and communication, the air system has also revolutionized our lives and our world. The major problem with the air system at the moment seems to be on the ground, where the growth of our airports is in direct conflict with the growth of our urban areas. We see the opportunity to alleviate this conflict in the very high speed ground systems such as maglev and anticipate that their first major application will be in enabling new forms of airports.

Will such systems spill over into the medium distance intercity market? Perhaps. While we still lack any operational experience with maglev, we would still prefer to bet on the adolescent technology of maglev rather than the very mature, almost geriatric, technology of rail. The extreme care rail requires the prevention of derailments, and the difficulty of containing the damage if it does occur bothers us; there are other approaches to guideway design that can virtually eliminate this concern.

The seeming inexorable push for personal services complicates imagining the future. The growth of air services has claimed increasing percentages of intercity passenger travel, but failure to cope with the problems engendered by its growth may slow its claiming of market share. There is no way to know.

What do we anticipate for our cities in the future? We have repeatedly made the point that the city is always a work in process, continuously evolving, but probably on a faster schedule of change and adaptation than ever before in history. There are many visions and scenarios [2], most of which, we suspect, may contain some truth, and none of them all of it. Our guess for the future is still more variety, slow but almost continuous evolution, leaving some parts untouched, some metamorphosed, some perhaps into enclaves unfamiliar to us today. We doubt that any single pattern will emerge, nor will any single interest group get their uncompromised dream city.

We have steered clear of the pejorative word sprawl because of the images attached to it. Sprawl means different things to different people. A critic might say that landscapes it is fashionable not to like such as low- and medium-income Los Angeles suburbs represent sprawl yet deny that George Washington's Mount Vernon overlooking the Capital represents sprawl. The sprawl debate pits central city property owners against developers on the fringe of the city. It pits those who see denser central city living as environmentally desirable against those who want less density, more greenery, and a less harsh imprint of structures on the land.

As we see it, there may indeed be more spreading out tomorrow. That would result from a more affluent population selecting the options for life and living that improved transportation enables. There might be more interaction with open space and natural habitat. Would that be terrible? Depends on how it is done. Surely we have the wit to create transportation, housing, water, sewage, and other designs that tread lightly on the land—design with nature, so to speak.

With respect to the expansion of urban areas, the rational Martian might point out that small numbers are involved. Land in urban uses is a fraction of a percentage of the land in the United States, and decades of continued urban expansion won't change that fraction very much. In response, the Venusian might rebut that urban expansion falls on nearby land that is high in precious open-space values. Perhaps better rural transportation services would ease this impact by vastly increasing the availability of accessible land and enable recreational and other activities to tread more lightly.

Returning to transportation options for the future, there is a two-way street. Not only will improving transportation spread into improvements in other processes, but the growth of new industries and consumption patterns; changes in settlement forms; mass migrations; pandemics; and tidal waves of developments in the society, the economy, or the polity will also drive transportation.

Thus the motivations for change in transportation are powerful, and we believe that the prospects are good for both better transportation and for its fruitful exploitation by our society and economy. Throughout our history our experiences in industry, education, and government exemplify a spirit of creativity, curiosity, and action that is almost unique in the world. Nothing in our more recent past suggests anything but a quickening of that spirit. We are blessed with both a culture and an economic system that encourages innovation. There will be potholes, but we are optimistic about where the road is going.

If change in transportation is a near-certainty, a second one is that the thoughts put forward in this book do not nearly exhaust the possibilities. New ideas and the entrepreneurship to make many of them reality will come from everywhere. And we encourage the reader to contribute to the supply.

It is fashionable to ask for specific performance goals. We have avoided this, treating the goal as just an increase in choices. To us, a path language seems useful because we cannot know the future well enough to use the language of specific goals. We have been thinking of ways the modes may be improved, primarily thinking in terms of development paths leading to an increased variety of services and production and consumption choices. We hope we have thought usefully about ways to proceed that make us better off, that might make our lives richer.

And we hope that you have found this journey both stimulating and enjoyable.

References

[1] National Safety Council, *Accident Facts, 1998 Edition*, Itaska, IL, 1998.

[2] Mathews, A., "2050: A Place Odyssey," *Preservation*, Vol. 51, No. 5, Sept./Oct. 1999, pp. 62–67.

About the Authors

William L. Garrison is Professor Emeritus of Civil and Environmental Engineering at the University of California at Berkeley. He is also Emeritus Research Engineer in the Institute for Transportation Studies at Berkeley, and is a past director of that institute. He remains active in teaching and research programs at the university.

Garrison received his Ph.D. degree from Northwestern University in 1950 and was on the faculty of several universities before moving to Berkeley. Over the years, he has served on many science and engineering advisory committees, including committees of the National Science Foundation, the Federal Highway and Transit Administrations, and the National Research Council.

He served as chairman of the Transportation Board of the council, and has participated in many other of the board's activities. Currently he serves on the board's Advisory Committee for the National Cooperative Research Program's investigations of new intelligent transportation systems concepts (IDEAS program).

Garrison's chief interests are in technological change in large systems and the social benefits of transportation investments. As is true of most vintage faculty, Garrison has produced a large number of papers and monographs. He has directed numerous Ph.D. theses, served on editorial advisory boards, and presented invited lectures at many universities. He has also worked with major corporations in the transportation industries.

Jerry D. Ward has been director of R&D policy in the U.S. DOT, senior associate in transportation in the Congress' Office of Technology Assessment, a visiting senior lecturer in transportation at MIT, and instigator and leader of much of the transportation research at Rockwell International.

Ward was active in the Transportation Research Board and in collaboration with the National Research Council's Advisory Committee on Transportation (where he met Bill Garrison). As a member of the new ITS America organization, he was a primary participant in instigating and defining the Intelligent Transportation System Architecture Synthesis program and was a member of U.S. DOT's Technical Review Team during the execution of that program. His most recent work was the formulation of strategies for urban congestion alleviation for the DOT.

He has written extensively about transportation for both technical and popular consumption.

Transportation is Ward's second career. After receiving a B.S. in physics from Caltech, he joined North American Aviation, where his forte turned out to be the analysis and synthesis of large-scale systems and their operational concepts, a background that turned out to be highly relevant to his later work in transportation systems. He became a vice president in the aircraft division.

But after 25 years in defense work the fun was gone. Supersaturated, Ward took a year's sabbatical, during which he earned an M.S. in business economics from UCLA. In 1972 he joined the U.S. DOT, launching his career in transportation.

Ward is now in active retirement and living in San Diego.

Index

311

Recent Titles in the Artech House ITS Library

John Walker, Series Editor

Advances in Mobile Information Systems, John Walker, editor

Incident Management in Intelligent Transportation Systems, Kaan Ozbay and Pushkin Kachroo

Intelligent Transportation Systems Architectures, Bob McQueen and Judy McQueen

ITS Handbook 2000: Recommendations from the World Road Association (PIARC), PIARC Committee on Intelligent Transport (Edited by Kan Chen and John C. Miles)

Positioning Systems in Intelligent Transportation Systems, Chris Drane and Chris Rizos

Smart Highways, Smart Cars, Richard Whelan

Tomorrow's Transportation: Changing Cities, Economies, and Lives, William L. Garrison and Jerry D. Ward

Vehicle Location and Navigation Systems, Yilin Zhao

Wireless Communications for Intelligent Transportation Systems, Scott D. Elliott and Daniel J. Dailey

For further information on these and other Artech House titles, including previously considered out-of-print books now available through our In-Print-Forever® (IPF®) program, contact:

Artech House
685 Canton Street
Norwood, MA 02062
Phone: 781-769-9750
Fax: 781-769-6334
e-mail: artech@artechhouse.com

Artech House
46 Gillingham Street
London SW1V 1AH UK
Phone: +44 (0)20 7596-8750
Fax: +44 (0)20 7630 0166
e-mail: artech-uk@artechhouse.com

Find us on the World Wide Web at:
www.artechhouse.com